Multilevel Statistical Models

Multilevel Statistical Models

Editor

Giorgos Michel

scitus
academics

Multilevel Statistical Models
Edited by **Giorgos Michel**

ISBN: 978-1-68117-446-4
Library of Congress Control Number: 2016937607

www.scitusacademics.com
Box No. 4766, 616 Corporate Way,
Suite 2, Valley Cottage,
NY 10989

Notice

Reasonable efforts have been made to publish reliable data and views articulated in the chapters are those of the individual contributors, and not necessarily those of the editors or publishers. Editors or publishers are not responsible for the accuracy of the information in the published chapters or consequences of their use. The publisher believes no responsibility for any damage or grievance to the persons or property arising out of the use of any materials, instructions, methods or thoughts in the book. The editors and the publisher have attempted to trace the copyright holders of all material reproduced in this publication and apologize to copyright holders if permission has not been obtained. If any copyright holder has not been acknowledged, please write to us so we may rectify.

Printed in United States of America on Acid Free Paper

Preface

In social, medical and other sciences, the importance of understanding complex hierarchical data structures is well understood. Multilevel modeling is now the accepted statistical technique for handling such information and is widely available in computer software packages. Therefore, a thorough understanding of these techniques is important for everyone working in these areas. Multilevel models (including linear hierarchical models, embedded models, mixed models, random effect models, random parameter models or split-plot designs) are statistical model parameters of more than one level. A student performance model that includes measurements for individual students and measures for classrooms in which students are grouped could be an example of this. These models can be seen as generalizations of linear models (particularly linear regression), but can also be extended to non-linear models. These models became much more popular after adequate computer power and software had been available. Multilevel models are especially suited for research designs in which data are organized at more than one level (i.e. Participants' nested data. The dependent variable must be examined at the lowest level of analysis.

The book combines the basic ideas of these techniques and shows how more complex models are derived. This book is an extensive assisting tool for graduate students. This book will benefit applied social science, economics, and biology and medicine statisticians.

Table of Contents

Chapter 1 A Bayesian Network View on Nested Effects Models 1

Chapter 2 Nested multiple imputation in large-scale assessments 27

Chapter 3 Spatial Random Slope Multilevel Modeling Using Multivariate Conditional Autoregressive Models: A Case Study of Subjective Travel Satisfaction in Beijing 67

Chapter 4 Dirichlet Compound Multinomials Statistical Models 113

Chapter 5 Modeling individual differences in randomized experiments using growth models: Recommendations for design, statistical analysis and reporting of results of internet interventions 141

Chapter 6 Studies of Climate Change with Statistical-Dynamical Models: A Review 189

Chapter 7 **Statistical Models for Web Pages Usability** **217**

Chapter 8 **Code-reading support environment visualizing three fields and educational practice to understand nested loops** **249**

Index **289**

CHAPTER 1

A Bayesian Network View on Nested Effects Models

Cordula Zeller,[1] Holger Fr¨ohlich,[2] and AchimTresch[3]

[1]Department of Mathematics, Johannes Gutenberg University, 55099 Mainz, Germany

[2]Division of Molecular Genome Analysis, German Cancer Research Center, 69120 Heidelberg, Germany

[3]Gene Center, Ludwig Maximilians University, 81377 Munich, Germany

ABSTRACT

Nested effects models (NEMs) are a class of probabilistic models that were designed to reconstruct a hidden signalling structure from a large set of observable effects caused by active interventions into the signalling pathway. We give a more flexible formulation of NEMs in the language of Bayesian networks. Our framework constitutes a natural generalization of the original NEM model, since it explicitly states the assumptions that are tacitly underlying the original version. Our approach gives rise to new learning methods for NEMs, which have been implemented in the R/*Bioconductor* package nem. We validate these methods in a simulation study and apply them to a synthetic lethality dataset in yeast.

1. INTRODUCTION

Nested effects models (NEMs) are a class of probabilistic models. They aim to reconstruct a hidden signalling structure (e.g., a gene regulatory system) by the analysis of high-dimensional phenotypes (e.g., gene expression profiles) which are consequences of well-defined perturbations of the system (e.g., RNA interference). NEMs have been introduced by Markowetz et al. [1], and they have been extended by Fröhlich et al. [2] and Tresch and Markowetz [3], see also the review of Markowetz and Spang [4]. There is an open-source software package "*nem*" available on the platform R/*Bioconductor* [5, 13], which implements a collection of methods for learning NEMs from experimental data. The utility of NEMs has been shown in several biological applications (*Drosophila melanogaster* [1], *Saccharomyces cerevisiae* [6], estrogen receptor pathway, [7]). The model in its original formulation suffers from some ad hoc restrictions which seemingly are only imposed for the sake of computability. The present paper gives an NEM formulation in the context of Bayesian networks (BNs). Doing so, we provide a motivation for these restrictions by explicitly stating prior assumptions that are inherent to the original formulation. This leads to a natural and meaningful generalization of the NEM model.

The paper is organized as follows. Section 2 briefly recalls the original formulation of NEMs. Section 3 defines NEMs as a special instance of Bayesian networks. In Section 4, we show that this definition is equivalent to the original one if we impose suitable structural constraints. Section 5 exploits the BN framework to shed light onto the learning problem for NEMs. We propose a new approach to parameter learning, and we introduce structure priors that lead to the classical NEM as a limit case. In Section 6, a simulation study compares the performance of our approach to other implementations. Section 7 provides an application of NEMs to synthetic lethality data. In Section 8, we conclude with an outlook on further issues in NEM learning.

2. THE CLASSICAL FORMULATION OF NESTED EFFECTS MODELS

For the sake of self-containedness, we briefly recall the idea and the original definition of NEMs, as given in [3]. NEMs are models that primarily intend to establish causal relations between a set of binary variables, the signals $\mathcal{S}$. The signals are not observed directly rather than through their consequences on another set of binary variables, the effects $\mathcal{E}$. A variable assuming the value 1, respectively, 0 is called *active*, respectively, *inactive*. NEMs deterministically predict the states of the effects, given the states of the signals. Furthermore, they provide a probabilistic model for relating the predicted state of an effect to its measurements. NEMs consist of a directed graph $\mathcal{T}$ the nodes of which are the variables $\mathcal{S} \cup \mathcal{E}$. Edges represent dependencies between their adjacent nodes. An arrow pointing from a to b means that b is active whenever a is active. To be more precise, the graph $\mathcal{T}$ can be decomposed into a graph Γ, which encodes the information flow between the signals, and a graph Θ which relates each effect to exactly one signal, see Figure 1. The effects that are active as a consequence of a signal s are those effects that can be reached from s via at most one step in Γ, followed by one step in Θ. Let $\delta_{s,e}$ denote the predicted state of e when signal s is activated, and let $\Delta = (\delta_{s,e})$ be the matrix of all predicted effects.

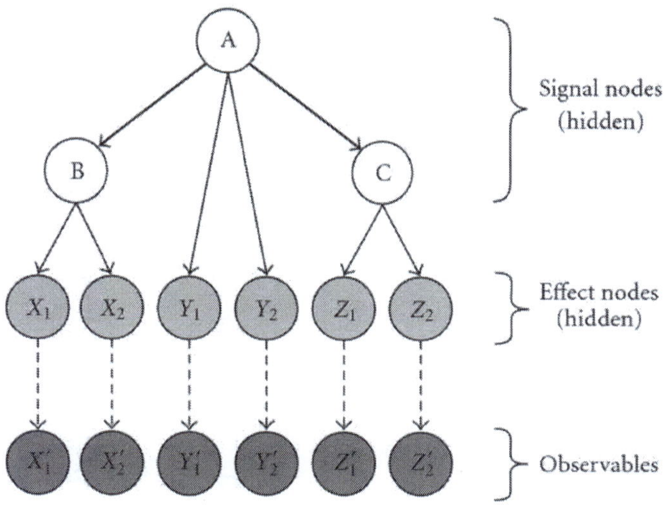

Figure 1. Example Of A Nested Effects Model In Its Bayesian Network Formulation. The bold arrows determine the graph Γ, the solid thin arrows encode Θ. Dashed arrows connect the effects to their reporters.

For the probabilistic part of the model, let $d_{s,e}$ be the data observed at effect e when signal s is activated (which, by the way, need not be binary and may comprise replicate measurements), and let $D = (d_{s,e})$ be the matrix of all measurements. The stochastic model that relates the predictions Δ to the experimental data D is given by a set of "local" probabilities $\mathcal{L} = \{p(d_{s,e} \mid e = \delta_{s,e}),\ s \in \mathcal{S}, e \in \mathcal{E}\}$. There are several ways of specifying $\mathcal{L}$, depending on the kind of data and the estimation approach one wants to pursue (see [1–3]). An NEM is completely parameterized by $\mathcal{T}$ and $\mathcal{L}$, and, assuming data independence, its likelihood is given by

$$p(D \mid \mathcal{T}, \mathcal{L}) = \prod_{s \in \mathcal{S}, e \in \mathcal{E}} p(d_{s,e} \mid e = \delta_{s,e}). \qquad (1)$$

3. THE BAYESIAN NETWORK FORMULATION OF NESTED EFFECTS MODELS

A Bayesian network describes the joint probability distribution of a finite family of random variables (the nodes) by a directed acyclic graph $\mathcal{T}$ and by a family of local probability distributions, which we assume to be parameterized by a set of parameters $\mathcal{L}$(for details, see, e.g., [8]). We want to cast the situation of Section 2 in the language of Bayesian networks. Assuming the acyclicity of the graph Γ of the previous section, this is fairly easy. A discussion on how to proceed when Γ contains cycles is given in Section 4. We have to model a deterministic signalling hierarchy, in which some components $(\mathcal{E})$can be probed by measurements, and some components $(\mathcal{S})$are perturbed in order to measure the reaction of the system as a whole. All these components $\mathcal{H} = \mathcal{S} \cup \mathcal{E}$will be *hidden* nodes in the sense that no observations will be available for $\mathcal{H}$, and we let the topology between these nodes be identical to that in the classical model. In order to account for the data, we introduce an additional layer of observable variables (*observables*, $\mathcal{O}$) in an obvious way: each effect node $e \in \mathcal{E}$has an edge pointing to a unique (its) observable node $e' \in \mathcal{O}$(see Figure 1). Hence, $\mathcal{O} = \{e' \mid e \in \mathcal{E}\}$, and we call e'*the observation of* e.

Let $pa(x)$be the set of parents of a node x, that is, the set of nodes that are direct predecessors of x. For notational convenience, we add a zero node z, $p(z = 0) = 1$, which has no parents, and which is a parent of all hidden nodes (but not of the observables). Note that by construction, $pa(x)$is not empty unless $x$is the zero node. For the hidden nodes, let the local probabilities describe a deterministic relationship,

$$p(x=1 \mid pa(x)) = \begin{cases} 1, & \text{if any parent of } x \text{ is active,} \\ 0, & \text{otherwise,} \end{cases}$$
$$= \max(pa(x)) \quad \text{for } x \in \mathcal{H}. \tag{2}$$

We slightly abuse notation by writing $\max(pa(x))$ for the maximum value that is assumed by a node in $pa(x)$. Obviously, all hidden nodes are set to 0 or 1 deterministically, given their parents. The local probabilities $p(e' \mid e)$, $e \in \mathcal{E}$, remain arbitrary for the moment. Assume that we have made an intervention into the system by activating a set of nodes $\mathcal{I} \subset \mathcal{S}$. This amounts to cutting all edges that lead to the nodes in $\mathcal{I}$ and setting their states to value 1. When an intervention $\mathcal{I}$ is performed, let $\delta_{\mathcal{I},h} \in \{0,1\}$ be the value of $h \in \mathcal{H}$. This value is uniquely determined by $\mathcal{I}$, as the next lemma shows.

Lemma 3.1.

$\delta_{\mathcal{I},h} = 1$ if and only if h can be reached from one of the nodes in $\mathcal{I}$ by a directed path in $\mathcal{T}$ (i.e., there exists a sequence of directed edges in $\mathcal{T}$ possibly of length zero, that links an $s \in \mathcal{I}$ to h). When performing an intervention $\mathcal{I}$, we, therefore, have

$$p(h=1) = \delta_{\mathcal{I},h} \quad \text{for } h \in \mathcal{H}. \tag{3}$$

Proof.

The proof is straightforward though somewhat technical and may be skipped for first reading. Let $\mathcal{H} = \{h_1, \ldots, h_n\}$ be an ordering of the nodes compatible with $\mathcal{T}$, which means $pa(h_j) \subseteq \{h_1, \ldots, h_{j-1}\}$,

$j = 1, \ldots, n$. Such an ordering exists because the graph connecting the states is acyclic. The proof is by induction on the order, the case $p(h_1 = 1) = \delta_{\mathcal{I},h_1}$ being trivial. If $h_j \in \mathcal{I}$, there is nothing to prove. Hence, we may assume $pa(h_j) \neq \varnothing$ in the graph which arises from $\mathcal{I}$ by cutting all edges that lead to a node in $\mathcal{I}$. Since $p(h_j = 1) = \max(pa(h_j))$, it follows that $\delta_{\mathcal{I},h_j} = 1$ if and only if $h_k = 1$ for some $h_k \in pa(h_j)$. This holds exactly if $\delta_{\mathcal{I},h_k} = 1$ for some $k \in pa(h_j)$ (in particular, $k < j$). By induction, this is the case if and only if there exists an $h_i \in \mathcal{I}$ and a directed path from h_i to h_k, which can then be extended to a path from h_i to h_j.

Let $D_{\mathcal{I}} = (e' = d_{e',\mathcal{I}}; e \in \mathcal{E})$ be an observation of the effects generated during intervention $\mathcal{I}$. Marginalization over the hidden nodes yields

$$p_{BN}(D_{\mathcal{I}}) = \sum_{(b_h) \in \{0,1\}^{\mathcal{H}}} P(D \mid h = b_h; h \in \mathcal{H}) \cdot P(h = b_h; h \in \mathcal{H}) . \quad (4)$$

Since by (3) there is only one possible configuration for the hidden nodes, namely, $s = \delta_{\mathcal{I},s}, S \in \mathcal{S}$, (4) simplifies to

$$\begin{aligned} p_{BN}(D_{\mathcal{I}}) &= P(D_{\mathcal{I}} \mid h = \delta_{\mathcal{I},h}; h \in \mathcal{H}) \\ &= P(D_{\mathcal{I}} \mid e = \delta_{\mathcal{I},e}; e \in \mathcal{E}) \end{aligned} \quad (5)$$

$$= \prod_{e \in \mathcal{E}} p(e' = d_{e',\mathcal{I}} \mid e = \delta_{\mathcal{I},e}) . \quad (6)$$

This formula is very intuitive. It says that if an intervention $\mathcal{I}$ has been performed, one has to determine the unique current state of each effect node. This, in turn, determines the (conditional) probability distribution of the corresponding observable node, for which one has to calculate the probability of observing the data. The product over all effects then gives the desired result.

4. SPECIALIZATION TO THE ORIGINAL NEM FORMULATION

In fact, (6) can be written as

$$
\begin{aligned}
P_{BN}(D_{\mathcal{I}}) &= \prod_{e \in \mathcal{E} \mid \delta_{\mathcal{I},e}=1} p(e' = d_{e',\mathcal{I}} \mid e = 1) \\
&\cdot \prod_{e \in \mathcal{E} \mid \delta_{\mathcal{I},e}=0} p(e' = d_{e',\mathcal{I}} \mid e = 0) \\
&= \prod_{e \in \mathcal{E} \mid \delta_{\mathcal{I},e}=1} \frac{p(e' = d_{e',\mathcal{I}} \mid e = 1)}{p(e' = d_{e',\mathcal{I}} \mid e = 0)} \\
&\cdot \prod_{e \in \mathcal{E}} p(e' = d_{e',\mathcal{I}} \mid e = 0) .
\end{aligned}
\quad (7)
$$

Let $r_{e,\mathcal{I}} = \log(p(e' = d_{e',\mathcal{I}} \mid e = 1) / p(e' = d_{e',\mathcal{I}} \mid e = 0))$, $e \in \mathcal{E}$, and $t_{\mathcal{I}} = \log \prod_{e \in \mathcal{E}} p(e' = d_{e',\mathcal{I}} \mid e = 0)$. Following the NEM formulation of [3], we consider all replicate measurements of an intervention $\mathcal{I}$ as generated from its own Bayesian network, and we try to learn the ratio $r_{e,\mathcal{I}}$ separately for each intervention $\mathcal{I}$. Therefore, we include $\mathcal{I}$ into the subscript. Taking logs in (7), it follows that

$$\log P_{BN}(D_{\mathcal{I}}) = \sum_{e \in \mathcal{E} | \delta_{\mathcal{I},e}=1} r_{e,\mathcal{I}} + t_{\mathcal{I}} = \sum_{e \in \mathcal{E}} \delta_{\mathcal{I},e} \cdot r_{e,\mathcal{I}} + t_{\mathcal{I}}. \quad (8)$$

Suppose that we have performed a series $\mathcal{I}_1, \ldots, \mathcal{I}_N \subseteq \mathcal{S}$ of interventions, and we have generated observations $D_1, \ldots, D_N$, respectively. Assuming observational independence, we get

$$\begin{aligned} \log P_{BN}(D_1, \ldots, D_N) &= \sum_{j=1}^{N} \log P(D_j) \\ &= \sum_{j=1}^{N} \sum_{e \in \mathcal{E}} \delta_{\mathcal{I}_j,e} \cdot r_{e,\mathcal{I}_j} + \sum_{j=1}^{N} t_{\mathcal{I}_j} \\ &= \sum_{j=1}^{N} (\Delta R)_{j,j} + \sum_{j=1}^{N} t_{\mathcal{I}_j} \\ &= \mathrm{tr}(\Delta R) + \sum_{j=1}^{N} t_{\mathcal{I}_j}, \end{aligned} \quad (9)$$

with the matrices $\Delta = (\delta_{\mathcal{I}_j,e})_{j,e}$ and $R = (r_{e,\mathcal{I}_j})_{e,j}$. The importance of (9) lies in the fact that it completely separates the estimation steps for $\mathcal{L}$ and $\mathcal{T}$. The information about the topology $\mathcal{T}$ of the Bayesian network enters the formula merely in the shape of Δ, and the local probability distributions alone define R. Hence, prior to learning the topology, one needs to learn the local probabilities only for once. Then, finding a Bayesian network that fits the data well means finding a topology which maximizes $\mathrm{tr}(\Delta R)$.

In the original formulation of NEMs, it is assumed that the set of interventions equals the set of all single-node interventions, $\mathcal{I}_s = \{s\}$, $s \in \mathcal{S}$. As pointed out in Section 2, the topology of the BN can be captured by two graphs Γ and Θ, which we identify with their corresponding adjacency matrices Γ and Θ by abuse of notation. The $\mathcal{S} \times \mathcal{S}$ adjacency matrix $\Gamma = (\Gamma_{s,t})_{s,t \in \mathcal{S}}$ describes the connections among signals, and the $\mathcal{S} \times \mathcal{E}$ adjacency matrix $\Theta = (\Theta_{s,e})_{s \in \mathcal{S}, e \in \mathcal{E}}$ encodes the connection between signals and effects. For convenience, let the diagonal elements of Γ equal 1. Denote by $\bar{\Gamma}$ the adjacency matrix of the transitive closure of Γ. Check that by Lemma 3.1, $\Delta = \bar{\Gamma}\Theta$. Therefore, we seek

$$\underset{(\Gamma,\Theta);\ \Gamma \text{acyclic}}{\arg\max} \ \operatorname{tr}(\bar{\Gamma}\Theta R), \tag{10}$$

which for transitively closed graphs $\Gamma = \bar{\Gamma}$ is exactly the formulation in [3]. It has the advantage that given $\bar{\Gamma}$, the optimal Θ can be calculated exactly and very fast, which dramatically reduces the search space and simplifies the search for a good graph $\bar{\bar{\Gamma}}$. The BN formulation of NEMs implies via (10) that two graphs Γ_1, Γ_2 are indistinguishable (likelihood equivalent, they fit all data equally well) if they have the same transitive closure. It is a subject of discussion whether the transitive closure of the underlying graph is a desirable property of such a model (think of causal chains which are observed in a stable state) or not (think of the dampening of a signal when passed from one node to another, or of a snapshot of the system where the signalling happens with large time lags), see [9].

It should be mentioned that the graph topology in our BN formulation of NEMs is necessarily acyclic, whereas the original formulation admits arbitrary graphs. This is only an apparent restriction. Due to the transitivity assumption, effects that connect to a cycle of signals will always react in the same way. This behaviour can also be obtained by arranging the nodes of the cycle in a chain and connecting the effects to the last node of the chain. This even leaves the possibility for connecting other effects to only a subset of the signals in the cycle by attaching them to a node higher up in the chain. As a consequence, admitting cycles does not extend the model class of NEMs in the Bayesian setting.

Although the original NEM model is algebraically and computationally appealing, it has some drawbacks. Learning the ratio $r_{e,\mathcal{I}} = \log(p(e' = d_{e',\mathcal{I}} \mid e = 1) / p(e' = d_{e',\mathcal{I}} \mid e = 0))$ separately for each intervention $\mathcal{I}$ entails various problems as follows.

1 Given an observation $d_{e'}$ at observable e' together with the state of its parent e, the quantity $p(e' = d_{e'} \mid e)$ should not depend on the intervention $\mathcal{I}$ during which the data were obtained, by the defining property of Bayesian networks. However, we learn the ratio $r_{e,\mathcal{I}}$ separately for each intervention, that is, we learn separate local parameters $\mathcal{L}$, which is counterintuitive.

2 Reference measurements $p(e' = d_{e',\mathcal{I}} \mid e = 0)$ are used to calculate the ratio $r_{e,\mathcal{I}}$, raising the need for a "null" experiment corresponding to an unperturbed observation $\mathcal{I}_0 = \varnothing$ of the system, which might not be available. The null experiment enters the estimation of each ratio $r_{e,\mathcal{I}}$. This introduces an unnecessary asymmetry in the importance of intervention $\mathcal{I}_0$ relative to the other interventions.

3 The procedure uses the data inefficiently since for a given topology, the quantities of interest $p(e' = d_{e'} \mid e = 1)$, respectively, $p(e' = d_{e'} \mid e = 0)$ could be learned from *all* interventions that imply $e = 1$, respectively, $e = 0$, providing a broader basis for the estimation.

The method proposed in the last item is much more time-consuming, since the occurring probabilities have to be estimated individually for each topology. However, such a model promises to better capture the real situation, so we develop the theory into this direction.

5. NEM LEARNING IN THE BAYESIAN NETWORK SETTING

Bear in mind that a Bayesian network is parameterized by its topology $\mathcal{T}$ and its local probability distributions, which we assume to be given by a set of local parameters $\mathcal{L}$. The ultimate goal is to maximize $P(\mathcal{T} \mid D)$. In the presence of prior knowledge, (we assume independent priors for the topology and the local parameters), we can write

$$\begin{aligned} P(\mathcal{T},\mathcal{L} \mid D) &= \frac{P(D \mid \mathcal{T},\mathcal{L})P(\mathcal{T},\mathcal{L})}{P(D)} \\ &\propto P(D \mid \mathcal{T},\mathcal{L})P(\mathcal{T})P(\mathcal{L}), \end{aligned} \tag{11}$$

from which it follows that

$$\begin{aligned} P(\mathcal{T} \mid D) &= \int P(\mathcal{T},\mathcal{L} \mid D)\mathrm{d}\mathcal{L} \\ &\propto P(\mathcal{T})\int P(D \mid \mathcal{T},\mathcal{L})P(\mathcal{L})\mathrm{d}\mathcal{L}. \end{aligned} \tag{12}$$

If it is possible to solve the integral in (12) analytically, it can then be used by standard optimization algorithms for the approximation of $\arg\max_{\mathcal{T}} P(\mathcal{T} \mid D)$. This full Bayesian approach will be pursued in Section 5.1. If the expression in (12) is computationally intractable or slow, we resort to a simultaneous maximum a posteriori estimation of $\mathcal{T}$ and $\mathcal{L}$, that is,

$$\begin{aligned}(\hat{\mathcal{T}},\hat{\mathcal{L}}) &= \arg\max_{\mathcal{T},\mathcal{L}} P(\mathcal{T},\mathcal{L} \mid D)\\ &= \arg\max_{\mathcal{T}}\left(\arg\max_{\mathcal{L}} P(D \mid \mathcal{T},\mathcal{L})P(\mathcal{L})\right)P(\mathcal{T}).\end{aligned} \quad (13)$$

The hope is that the maximization $\hat{\mathcal{L}}(\mathcal{T}) = \arg\max_{\mathcal{L}} P(D \mid \mathcal{T},\mathcal{L})P(\mathcal{L})$ in (13) can be calculated analytically or at least very efficiently, see [3]. Then, maximization over $\mathcal{T}$ is again done using standard optimization algorithms. Section 5.2 is devoted to this approach.

5.1. Bayesian Learning of the Local Parameters

Let the topology $\mathcal{T}$ and the interventions $\mathcal{I}_j$ be given. Let N_{eik} denote the number of times the observable e was reported to take the value k, while its true value was i, and let N_{ei} be the number of measurements taken from e when its true value is i:

$$\begin{aligned}N_{eik} &= \mid \{j \mid \delta_{\mathcal{I}_j,e} = i, d_{e',\mathcal{I}_j} = k\} \mid,\\ N_{ei} &= \mid \{j \mid \delta_{\mathcal{I}_j,e} = i\} \mid.\end{aligned} \quad (14)$$

Binary Observables

The full Bayesian approach in a multinomial setting was introduced by Cooper and Herskovits [10].

The priors are assumed to follow beta distributions:

$$\beta_0 \sim \mathrm{Beta}(\alpha_0, \beta_0), \qquad \beta_1 \sim \mathrm{Beta}(\alpha_1, \beta_1). \tag{15}$$

Here, $\alpha_0, \alpha_1, \beta_0$, and $\beta_1$are shape parameters, which, for the sake of simplicity, are set to the same value for every effect e. This assumption can be easily dropped and different priors may be used for each effect.

In this special setting with binomial nodes with one parent, the well-known formula of Cooper and Herskovitz can be simplified to

$$\begin{aligned} P(D_1, \ldots, D_N \mid \mathcal{T}) &= \prod_{j=1}^{N} \prod_{e \in \mathcal{E}} \prod_{i \in \{0,1\}} \frac{\Gamma(N_{ei0}+\alpha_i)\Gamma(N_{ei1}+\beta_i)\Gamma(\alpha_i+\beta_i)}{\Gamma(N_{ei}+\alpha_i+\beta_i)\Gamma(\alpha_i)\Gamma(\beta_i)} \\ &\propto \prod_{j=1}^{N} \prod_{e \in \mathcal{E}} \prod_{i \in \{0,1\}} \frac{\Gamma(N_{ei0}+\alpha_i)\Gamma(N_{ei1}+\beta_i)}{\Gamma(N_{ei}+\alpha_i+\beta_i)}. \end{aligned} \tag{16}$$

Continuous Observables

Let us assume $p(e' \mid e=k)$to be normally distributed with mean $a_{ek}$and variance σ^2_{ek}, $e \in \mathcal{E}$, $k \in \{0,1\}$. We refer to the work of

Neapolitan [8] for the calculation of this section. Let the prior for the precision $r_{ek} = 1/\sigma^2_{ek}$ follow a Gamma distribution,

$$p(r_{ek}) = \mathrm{Gamma}\left(r_{ek}; \frac{\alpha}{2}, \frac{\beta}{2}\right). \tag{17}$$

Given the precision r_{ek}, let the conditional prior for the mean a_{ek} be

$$p(a_{ek} \mid r_{ek}) = N\left(a_{ek}; \mu, \frac{1}{v r_{ek}}\right). \tag{18}$$

So the Data of observable e given its parent's state $\delta_{\mathcal{I}_j,e} = k$ is

$$p(d_{e',\mathcal{I}_j} \mid a_{ek}, r_{ek}) = N\left(d_{e',\mathcal{I}_j}; a_{ek}, \frac{1}{r_{ek}}\right), \quad \delta_{\mathcal{I}_j,e} = k. \tag{19}$$

Then,

$$\begin{aligned}
&P(D_1, \ldots, D_N \mid \mathcal{T}) \\
&= \prod_{e \in \mathcal{E}} \prod_{k \in \{0,1\}} \left(\frac{1}{2\pi}\right)^{N_{ek}/2} \left(\frac{v}{v+N_{ek}}\right)^{1/2} 2^{N_{ek}/2} \frac{\Gamma((\alpha+N_{ek})/2)}{\Gamma(\alpha/2)} \\
&\quad \cdot \frac{\beta|^{\alpha/2}}{|\beta+s_{ek}+(vN_{ek}/(v+N_{ek}))(\bar{x}_{ek}-\mu)^2|^{(\alpha+N_{ek})/2}} \\
&\propto \prod_{e \in \mathcal{E}} \prod_{k \in \{0,1\}} \left(\frac{v}{v+N_{ek}}\right)^{1/2} \\
&\quad \times \frac{\Gamma((\alpha+N_{ek})/2)}{|\beta+s_{ek}+(vN_{ek}/(v+N_{ek}))(\bar{x}_{ek}-\mu)^2|^{(\alpha+N_{ek})/2}}.
\end{aligned} \tag{20}$$

The data enters this equation via

$$\bar{x}_{ek} = \frac{1}{N_{ek}} \sum_{j \mid \delta_{\mathcal{I}_j, e} = k} d_{e', \mathcal{I}_j}, \qquad s_{ek} = \sum_{j \mid \delta_{\mathcal{I}_j, e} = k} (d_{e', \mathcal{I}_j} - \bar{x}_{ek})^2.$$

(21)

5.2. Maximum Likelihood Learning of the Local Parameters

Let the topology $\mathcal{T}$ and the interventions $\mathcal{I}_j$ be given. For learning the parameters of the local distributions $p(e' \mid e)$, we perform maximum likelihood estimation in two different settings. The observables are assumed to follow either a binomial distribution or a Gaussian distribution.

Binary Observables

For an effect $e \in \mathcal{E}$, let its observable e' be a binary random variable with values in $\{0, 1\}$, and let $p(e' = 1 \mid e = x) = \beta_{e,x}$, $x \in \{0, 1\}$. The model is then completely parameterized by the topology $\mathcal{T}$ and $\mathcal{L} = \{\beta_{e,x} \mid e \in \mathcal{E},\ x \in \{0, 1\}\}$.

Note that

$$
\begin{aligned}
&P(D_1,\ldots,D_N \mid \mathcal{T},\mathcal{L}) \\
&= \prod_{j=1}^{N} \prod_{e\in\mathcal{E}} p(e' = d_{e',\mathcal{I}_j} \mid e = \delta_{\mathcal{I}_j,e}) \\
&= \prod_{e\in\mathcal{E}} \prod_{x\in\{0,1\}} \prod_{j \mid \delta_{\mathcal{I}_j,e}=x} p(e' = d_{e',\mathcal{I}_j} \mid e = x) \\
&= \prod_{e\in\mathcal{E}} \prod_{x\in\{0,1\}} B(k = N_{ex1}; n = N_{ex}, p = \beta_{e,x}),
\end{aligned}
\qquad (22)
$$

with $B(k;n,p) = \binom{n}{k} p^k (1-p)^{n-k}$. The parameter set $\widehat{\mathcal{L}}$ that maximizes expression (22) is

$$
\widehat{\beta}_{e,x} = \frac{N_{ex1}}{N_{ex}}, \quad e \in \mathcal{E},\ x \in \{0,1\} \qquad (23)
$$

(the ratios with a denominator of zero are irrelevant for the evaluation of (22) and are set to zero).

Continuous Observables

There is an analogous way of doing ML estimation in the case of continuous observable variables if one assumes $p(e' \mid e = x)$ to be a normal distribution with mean $\mu_{e,x}$ and variance $\sigma^2_{e,x}$, $e \in \mathcal{E}$, $x \in \{0,1\}$.

Note that

$$\begin{aligned}
&P(D_1,\dots,D_N \mid \mathcal{T},\mathcal{L}) \\
&= \prod_{j=1}^{N} \prod_{e\in\mathcal{E}} p(e' = d_{e',\mathcal{I}_j} \mid e = \delta_{\mathcal{I}_j,e}), \\
&= \prod_{e\in\mathcal{E}} \prod_{x\in\{0,1\}} \prod_{j \mid \delta_{\mathcal{I}_j,e}=x} p(e' = d_{e',\mathcal{I}_j} \mid e = x) \\
&= \prod_{e\in\mathcal{E}} \prod_{x\in\{0,1\}} \mathcal{N}(\{d_{e',\mathcal{I}_j} \mid \delta_{\mathcal{I}_j,e} = x\};\ \mu_{e,x},\sigma_{e,x}),
\end{aligned} \tag{24}$$

With

$$\begin{aligned}
&\mathcal{N}(\{x_1,\dots,x_k\};\ \mu,\sigma) \\
&= \left(\frac{1}{(\sqrt{2\pi}\,\sigma)^k}\right)\cdot \exp\left(-\left(\sum_{j=1}^{k}\frac{(x_j-\mu)^2}{2\sigma^2}\right)\right).
\end{aligned} \tag{25}$$

The parameter set $\hat{\mathcal{L}}$ maximizing expression (24) is

$$\hat{\mu}_{e,x} = \frac{1}{N_{ex}} \sum_{j \mid \delta_{\mathcal{I}_j,e}=x} d_{e',\mathcal{I}_j},$$

$$\hat{\sigma}_{e,x} = \frac{1}{N_{ex}} \sum_{j \mid \delta_{\mathcal{I}_j,e}=x} (d_{e',\mathcal{I}_j} - \hat{\mu}_{e,x})^2, \quad e\in\mathcal{E},\ x\in\{0,1\} \tag{26}$$

(quotients with a denominator of zero are again irrelevant for the evaluation of (24) and are set to zero). Note that in both the discrete and the continuous case, $\tilde{\mathcal{L}}$ depends on the topology $\mathcal{T}$, since the topology determines the values of $\delta_{\mathcal{I}_j,e}$, $j = 1,\ldots,N, e \in \mathcal{E}$.

5.3. Structure Learning

It is a major achievement of NEMs to restrict the topology of the underlying graphical structure in a sensible yet highly efficient way, thus, tremendously reducing the size of the search space. There is an arbitrary "core" network consisting of signal nodes, and there is a very sparse "marginal" network connecting the signals to the effects. It is, however, by no means necessary that the core network and the signal nodes coincide. We propose another partition of the hidden nodes into core nodes $\mathcal{C}$ and marginal nodes $\mathcal{M}$, $\mathcal{H} = \mathcal{C} \cup \mathcal{M}$, which may be distinct from the partition into signals and effects, $\mathcal{H} = \mathcal{S} \cup \mathcal{E}$. No restrictions are imposed on the subgraph generated by the core nodes (except that the graph has to be acyclic). The key semantics of NEMs is that marginal nodes are viewed as the terminal nodes of a signalling cascade. The requirement that the marginal nodes have only few or at most one incoming edge can be translated into a well-known structure prior $P(\mathcal{T})$ (see, e.g., [12]) which penalizes the number of parents of marginal nodes:

$$\log P(\mathcal{T}) = -v \cdot \sum_{m \in \mathcal{M}} \max(|pa(m)| - 1, 0). \quad (27)$$

For the penalty parameter $v = \infty$, this is the original NEM restriction. If $v = 0$, each marginal node can be assigned to all suitable core nodes. As a consequence, there is always a best scoring topology with an empty core graph. νmakes signalling to the marginal nodes "expensive" relative to signalling in the core graph. It is unclear how to choose νoptimally, so we stick to the choice $\nu = \infty$for the applications. Simulation studies have shown that a simple gradient ascent algorithm does very well in optimizing the topology of the Bayesian network, compared to other methods that have been proposed [7].

6. SIMULATION

6.1. Network and Data Sampling

The ML and the Bayesian method for parameter learning have been implemented in the *nem* software [13], which is freely available at the R/*Bioconductor* software platform [5]. To test the performance of our method, we conducted simulations with randomly created acyclic networks with $n = 4$signals. The out-degree $d$of each signal was sampled from the power-law distribution

$$p(d) = \frac{1}{Z} d^{-2.5}, \tag{28}$$

where $Z$is an appropriate normalization constant. Binary data (1 = effect, 0 = no effect) was simulated for the perturbation of each signal in the created network using 4 replicate measurements with type-I and type-II error rates $\alpha$and β, which were drawn uniformly from $[0.1, 0.5]$and $[0.01, 0.2]$for each perturbation separately. This simulates individual measurement error characteristics for each experiment.

6.2. Results

We compared our Bayesian network model with the classical NEM using a greedy hill-climbing algorithm to find the best fitting connection between signals. We simulated $m = 25, 50, 100$ and 250 effect nodes, and for each number of effects, 100 random networks were created as described above. Figure 2 demonstrates that both approaches perform very similarly.

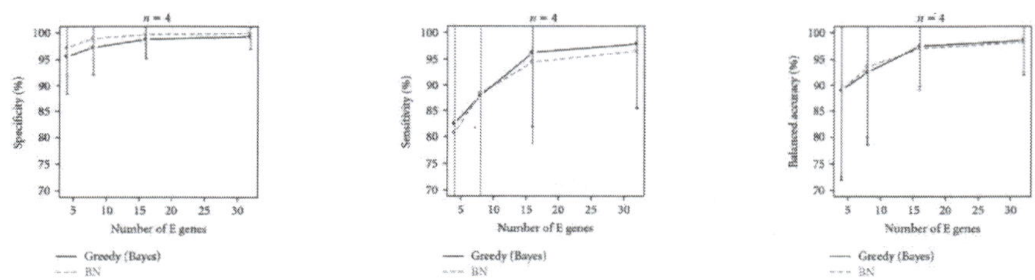

Figure 2 . Results (Specificity, Sensitivity, And Balanced Accuracy) Of Simulation Run. The continuous line (greedy (Bayes)) describes the performance of the traditional NEM method, the dashed line stands for our new approach via Bayesian networks.

7. APPLICATION

We apply the BN formulation of the NEM methodology to a dataset of synthetic lethality interactions in yeast. We reveal hierarchical dependencies of protein interactions. Synthetic lethality (SL) is the phenomenon that a cell survives the single gene deletion of a gene A and a gene B, but the double deletion of A and B is detrimental. In this case, A and B are called SL partners or an SL pair. It has been shown in [11] that it is not so much SL partners themselves whose gene products participate in the same protein complex or pathway, rather than genes that *share* many SL partners. The detection of genetic interactions via synthetic lethality screens and appropriate computational tools is a current area of research, see [14]. Ye and Peyser define a hypergeometric score function to test whether two genes have many SL partners in common. They apply their methodology to a large SL data set [15] for finding pairs (and, consequently, clusters) of genes whose products are likely to participate in the same pathway. We extend their approach as explained in Figure 3. SL partnership arises (not exclusively, but prevalently) among genes pertaining to two distinct pathways that complement each other in a vital cell function. If a gene A is upstream of gene B in some pathway, a deletion of gene A will affect at least as many pathways as a deletion of gene B. Hypothesizing a very simplistic world, all SL partners of B will as well be SL partners of A; but this subset relation can be detected by NEMs. Take the primary knockout genes as core nodes, and the secondary knockout genes as marginal nodes, which are active given a primary knockout whenever SL occurs. We used the dataset from [15] and chose 40 primary knockout genes having the most SL interaction partners as core genes, and included all their 194 SL partners as marginal nodes. An NEM with binary observables was estimated, both with the maximum likelihood approach and in the Bayesian setting. It should be emphasized that NEM estimation for this dataset is only possible in the new BN setting because there is no canonical "null experiment," which enables us to estimate the likelihood ratios $r_{I,e}$ needed in the classical setting in (7), (8), [14].

Figure 4 displays the results of the NEM reconstruction. The NEMs estimated by both methods agree well as far as the hierarchical organisation of the network is concerned. However, they do not agree well with the clusters found in [11]. We refrain from a biological interpretation of these networks, since the results are of a preliminary nature. In particular, the reconstruction does not take advantage of prior knowledge, and the postulated edges were not validated experimentally.

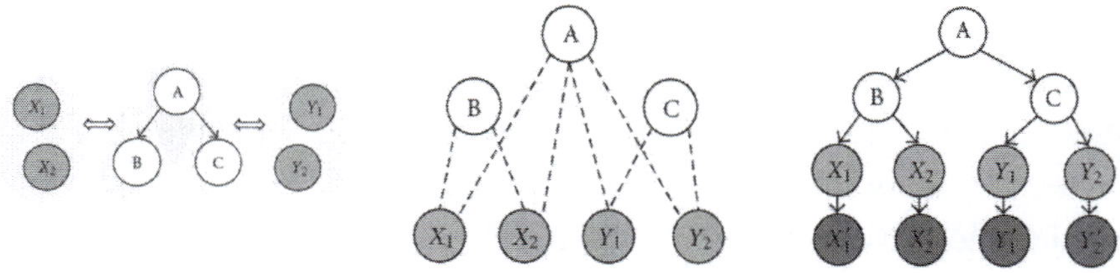

Figure 3. Schematic Reconstruction Of A Signalling Pathway Through Synthetic Lethality Data. (a) A situation in which there are two pairs of complementary pathways {A, B}, {X_1, X_2} and {A,C}, {Y_1,Y_2}). (b) Model of the situation as follows: the primary knockouts are considered signals {A, B, C} (they are not observed). As those are our genes of interest, they will also form the core nodes. The secondary effects are accessible to observation and, therefore, represented by the effects X_1, X_2, Y_1, and Y_2. Each SL pair is connected by a dashed line. (c) NEMs that might be estimated from (b), using binary observables and one of the approaches in Sections 5.1 or 5.2.

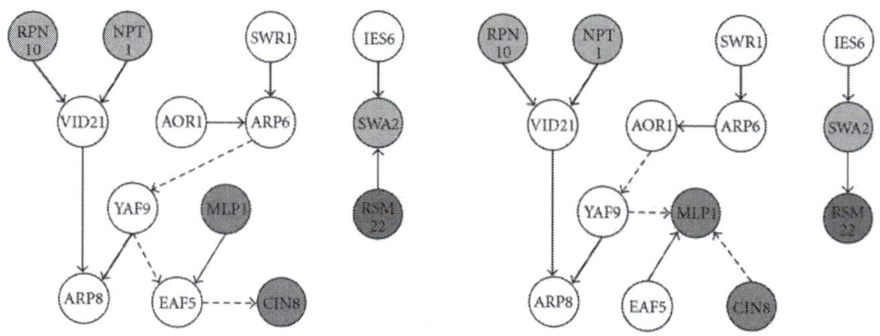

Figure 4 . Nems Constructed From The Sl Data. Only core genes that have at least one edge are shown. (a) The ML estimate. (b) The Bayesian estimate (the prior choice (see (15)) was β_{e0} ~Beta(5, 2), respectively, β_{e1} ~Beta(2, 5)). Nodes with the same shading pertain to the same clusters that were defined by Ye et al. [11]. Bold arrows appear in both reconstructions, thin arrows reverse their direction, and dashed arrows are unique to each reconstruction.

8. SUMMARY AND OUTLOOK

Some aspects of the classical NEM concept appear in a different light when stated in the BN framework. Mainly, these are three folds: (1) the learning of the local parameters, for which we proposed new learning rules; (2) the structural constraints, they can be cast as priors on the NEM topology; (3) the distinction between hidden and observable nodes, which can be different from that of core nodes and marginal nodes.

We proposed some new lines of investigation, like a full Bayesian approach for the evaluation of $P(T \mid D)$, and a smooth structure prior with continuous penalty parameter ν. It is much easier to proceed in the BN framework and implement, for example, a boolean logic for the signal transduction, which is less simplistic than in the current model. A straightforward application of NEMs in their BN formulation to synthetic lethality data demonstrated the potential of the NEM method, with the purpose of stimulating further research in that field.

ACKNOWLEDGMENTS

The authors like to thank Peter Bühlmann and Daniel Schöner for proposing the application of NEMs to synthetic lethality data. This work was supported by the Deutsche Forschungsgemeinschaft, the Sonderforschungsbereich SFB646. H. Fröhlich is funded by the National Genome Research Network (NGFN) of the German Federal Ministry of Education and Research (BMBF) through the platforms SMP Bioinformatics (OIGR0450) and SMP RNA (OIGR0418).

REFERENCES

1. Markowetz F, Bloch J, Spang R: Non-transcriptional pathway features reconstructed from secondary effects of RNA interference. Bioinformatics 2005, 21(21):4026-4032. 10.1093/bioinformatics/bti662

2. Fröhlich H, Fellmann M, Sültmann H, Poustka A, Beissbarth T: Estimating large-scale signaling networks through nested effect models with intervention effects from microarray data. Bioinformatics 2008, 24(22):2650-2656. 10.1093/bioinformatics/btm634

3. Tresch A, Markowetz F: Structure learning in nested effects models. Statistical Applications in Genetics and Molecular Biology 2008., 7(1, article 9):

4. Markowetz F, Spang R: Inferring cellular networks—a review. BMC Bioinformatics 2007, 8, supplement 6: 1-17.

5. Gentleman RC, Carey VJ, Bates DM, et al.: Bioconductor: open software development for computational biology and bioinformatics. Genome biology 2004, 5(10, article R80):1-16. 10.1186/gb-2004-5-10-r80

6. Markowetz F, Kostka D, Troyanskaya OG, Spang R: Nested effects models for high-dimensional phenotyping screens. Bioinformatics 2007, 23(13):i305-i312. 10.1093/bioinformatics/btm178

7. Froehlich H, Fellmann M, Sueltmann H, Poustka A, Beissbarth T: Large scale statistical inference of signaling pathways from RNAi and microarray data. BMC Bioinformatics 2007, 8, article 386: 1-15.

8. Neapolitan RE: Learning Bayesian Networks. Prentice Hall, Upper Saddle River, NJ, USA; 2003.

9. Jacob J, Jentsch M, Kostka D, Bentink S, Spang R: Detecting hierarchical structure in molecular characteristics of disease using transitive approximations of directed graphs. Bioinformatics 2008, 24(7):995-1001. 10.1093/bioinformatics/btn056

10. Cooper GF, Herskovits E: A Bayesian method for the induction of probabilistic networks from data. Machine Learning 1992, 9(4):309-347.

11. Ye P, Peyser BD, Pan X, Boeke JD, Spencer FA, Bader JS: Gene function prediction from congruent synthetic lethal interactions in yeast. Molecular Systems Biolog 2005, 1, article 2005.0026: 1. 10.1038/msb4100034

12. Mukherjee S, Speed TP: Network inference using informative priors. Proceedings of the National Academy of Sciences of the United States of America 2008, 105(38):14313-14318. 10.1073/pnas.0802272105

13. Fröhlich H, Beißbarth T, Tresch A, et al.: Analyzing gene perturbation screens with nested effects models in R and bioconductor. Bioinformatics 2008, 24(21):2549-2550. 10.1093/bioinformatics/btn446

14. Le Meur N, Gentleman R: Modeling synthetic lethality. Genome Biology 2008, 9(9, article R135):1-10. 10.1186/gb-2008-9-9-r135

15. Tong AHY, Lesage G, Bader GD, et al.: Global mapping of the yeast genetic interaction network. Science 2004, 303(5659):808-813. 10.1126/science.1091317

CHAPTER 2

Nested multiple imputation in large-scale assessments

Sebastian Weirich[1*], Nicole Haag[1], Martin Hecht[1], Katrin Böhme[1], Thilo Siegle[1] and Oliver Lüdtke[2]

[1]Institute for Educational Quality Improvement, Humboldt-Universität zu Berlin, Unter den Linden 6, 10099 Berlin,Germany.

[2]Leibniz Institute for Science and Mathematics Education (IPN), Centre for International Student Assessment, Olshausenstraße 62, 24118 Kiel, Germany.

ABSTRACT

Background

In order to measure the proficiency of person populations in various domains, large-scale assessments often use marginal maximum likelihood IRT models where person proficiency is modelled as a random variable. Thus, the model does not provide proficiency estimates for any single person. A popular approach to derive these proficiency estimates is the multiple imputation of plausible values (PV) to enable subsequent analyses on complete data sets. The main drawback is that all variables that are to be analyzed later have to be included in the imputation model to allow the distribution of plausible values to be conditional on these variables. These background variables (e.g., sex, age) have to be fully observed which is highly unlikely in practice. In several current large-scale assessment programs missing observations on background variables are dummy coded, and subsequently, dummy codes are used additionally in the PV imputation model. However, this approach is only appropriate for small proportions of missing data. Otherwise the resulting population scores may be biased.

Methods

Alternatively, single imputation or multiple imputation methods can be used to account for missing values on background variables. With both imputation methods, the result is a two-step procedure in which the PV imputation is nested within the background variable imputation. In the *single+multiple-imputation* (SMI), each missing value on background variables is replaced by one value. In *the multiple+multiple-imputation* (MMI), each missing value is replaced by a set of imputed values. MMI is expected to outperform SMI as SMI ignores the uncertainty due to missing values in the background data.

Results

In a simulation study, both methods yielded unbiased population estimates under most conditions. Still, the recovery proportion was slightly higher for the MMI method.

Conclusions

The advantages of the MMI method are apparent for fairly high proportions of missing values in combination with fairly high dependency between the latent trait and the probability of missing data on background variables.

KEYWORDS

Large-scale assessment Missing data Imputation Simulation Item response theory

1. BACKGROUND

Several large-scale assessment programs such as the National Assessment of Educational Progress (NAEP), the Trends in International Mathematics and Science Study (TIMSS), and the Programme for International Student Assessment (PISA) employ item response theory (IRT) methods to measure the achievement of populations of examinees in various domains, for example, reading and science, for the purpose of system monitoring. The central interest is not to obtain individual proficiency scores but to obtain population and subpopulation estimates of person proficiency, denoted by θ. In addition to the item responses, the estimation process includes background information for each person (often collected via questionnaires) comprising a multitude of variables, such as demographic, social, or motivational variables. Because these variables may be associated with θ, they are of particular importance in the estimation of subpopulation differences. Large-scale assessment programs are often confronted with the problem of missing values in item responses or in background data. Planned missing data in the item responses are a direct consequence of multiple matrix sampling designs (Frey et al. [2009]; Gonzalez & Rutkowski [2010]). Methods of item and person parameter estimation which incorporate missing data in the item responses have long been developed and have been proved to be suitable (Lord [1974], [1983]; Mislevy et al. [1992]). By contrast, a considerable amount of missing data in background variables may lead to biased estimates of population parameters (Rutkowski [2011]). Therefore, the present article examines the problem of missing data in background information when population estimates are of interest. We compared two methods that are based on imputation methods with regard to the bias and root mean square error (RMSE) of population parameters. The results support the implementation of multiple imputation methods in the treatment of missing data in background information.

1.1 Item and person parameter estimation

In large-scale assessments, the estimation of item and person parameters commonly uses the marginal maximum likelihood (MML) method (Embretson & Reise [2000]; Tuerlinckx et al. [2004]; Wilson & De Boeck [2004]). Conceptually, the estimation process is composed of two parts: the *item response model*, which results from an analysis of the response patterns, and the *population model*, which results from incorporating background information (Adams & Wu [2007]). von Davier et al. ([2009]) refer to these two parts as the likelihood function and the prior distribution function. A model that comprises both parts is called a *conditioning model* or combined model (Adams & Wu [2007]).

In the item response model, the categorical response patterns to a set of test items are modeled as the dependent variable in a logistic regression with item difficulties and person proficiencies as independent variables. Adams and Wu ([2007]) and Adams et al. ([1997]) described the model in its generalized form. For the purpose of illustration, we will use the special case of a unidimensional model with dichotomous items (Rasch model):

$$\text{logit}(P(X_{ni} = 1)) = \theta_n - b_i, \tag{1}$$

where X_{ni} represents the observed response of a person n for an item i, θ_n represents the proficiency score of person n, and b_i represents the difficulty of item i. The definition of the model in the MML formulation needs to specify a density for the latent variable θ (Adams & Wu [2007]). The default assumption is that persons are sampled from a normal population with mean μ and variance σ^2. Therefore, θ is a random effect, and the population model is: $\theta_n = \mu + E_n$, with $E_n \sim N(0, \sigma^2)$. If we assume that θ depends on background variables, the population model may be augmented to a regression model with several predictors (Adams & Wu [2007]):

$$\theta_n = \mathbf{Y_n}\boldsymbol{\beta} + E_n \tag{2}$$

where $\mathbf{Y}_N$ is a vector of u fixed and known values for a set of u variables related to each person, and $\mathbf{B}$ is the corresponding vector of regression coefficients to be estimated. E_n is assumed to be independently and identically normally distributed with mean zero and variance σ^2; that is, $E_n \sim N(0, \sigma^2)$. The population model now incorporates fixed and random effects and provides *a priori* information about the expected person distribution given all available background information (von Davier et al. [2009]).

The conditioning model is the item response model multiplied by the population model (Adams & Wu [2007]; Embretson & Reise [2000]). Hence, the conditioning model captures information from the test items (expressed in the item response model) and background information (expressed in the population model). Hence, it offers the opportunity to specify a population distribution of persons consisting of several subpopulations. This is especially important if the population distribution is conditional on several demographic variables rather than being a simple (unconditional) normal distribution.

1.2 Types of missing data and their effects on the estimation process

In large-scale assessments, three types of missing data can be distinguished. The first one is rather fundamental: IRT models are usually intended to measure proficiencies of persons. As each person's individual proficiency θ_n is considered to be a latent construct, θ is an inherently unobserved (i.e., missing) variable. Missing data in the individual response pattern (the 0/1 answers of persons to test items) constitute the second type, and missing data in background information constitute the third type of missing data.

1.3 First type of missing data: inherently missing θ

The inherently missing person parameter θ_n is not a problem in MML estimation. The MML formulation of the Rasch model may be described as a random person–fixed item model (De Boeck [2008]) and does not parametrize individual θ_n parameters. However, a common practice in most large-scale assessment programs is to provide individual θ_n values for subsequent analyses (e.g., ordinary least squares regression, contingency table analyses). θ_n estimates therefore have to be generated in an additional step, using the parameters of the fitted model. The most suitable method to generate θ_n estimates is plausible value imputation (Mislevy et al. [1992]; von Davier et al. [2009]; Wu [2005]). This method is based on multiple imputation (Rubin [1987]): The conditioning model defines how the observed item responses are related to θ, and the plausible-values approach provides a model-based imputation of missing θ_n values. Mislevy et al. ([1992]) pointed out that plausible values provide synthetic data produced by the model. Analyzing plausible values with standard techniques yields the same results as if we actually observed true person proficiencies (Rubin [1987]). This does not necessarily hold for other estimates of individual θ_n values, for example Weighted Maximum Likelihood Estimates (WLEs; Warm [1989]) or the expected a posteriori distribution (EAP; Bock & Aitkin [1981]). Moreover, the *multiple* imputation of plausible values allows investigators to adequately quantify the uncertainty in the parameter estimation of subsequent analyses.

The imputation of plausible values can be conditioned on background information by including some demographic variables when specifying the population model. Similar to the recommendations made for multiple imputation (Rubin [1987]), it is crucial to include all variables that may be used in subsequent analyses in order to ensure unbiased estimation of their effects on achievement (Mislevy et al. [1992]; von Davier et al. [2009]).

1.4. Second type of missing data: missing data in the item responses

The second (and also the third) type of missing data concerns missing values in data that are expected to be observed. However, the second type of missing data likewise is not a problem for parameter estimation, if some conditions are met. Missing data in the item responses are a direct consequence of multiple matrix sampling designs (Frey et al., [2009]; Gonzalez & Rutkowski [2010]). As large samples of items are used to comprehensively cover the test constructs, only a subset of items is presented to each person to keep the individual workload within acceptable boundaries. Such missing responses which result from giving samples of items to samples of persons are often called *missing by design* (Enders [2010]). If the test booklets are randomly distributed to the persons, these data are assumed to be missing completely at random (MCAR). MCAR means that missing values on a variable neither depend on values of other variables in the model nor on values of the variable itself (Graham [2009]; Rubin [1987]). By contrast, responses that a person has chosen to omit are often treated as *incorrect* responses. See Lord ([1974], [1983]) and Ludlow and O'leary ([1999]) for a discussion of this procedure.

1.5. Third type of missing data: missing data in background information

The third type of missing data—which is the focus of the present article—concerns missing values on background variables which are inherently due neither to the latent variable modeling of θ nor to the test design. In the population model (Equation 2), it is assumed that background variables are measured without error because $\mathbf{Y}_N$ is a vector of fixed and known values (Adams & Wu [2007]). In principle, the same problem applies when missing values occur in ordinary least squares regression models (e.g., Little [1992]): The corresponding procedures require that all variables be fully observed.

What might be the consequences of missing values in background data? In theory, two possible effects would be conceivable: First, the parameter estimates in Equation 2 might be biased which may, in turn, result in biased estimates of subpopulation differences. Thus, the effect of missing data on the population model is comparable to the effect of missing data on ordinary least-squares regression models. Second, the uncertainty of the corresponding mean estimates might not be adequately represented which may result in biased standard errors of the subpopulation mean estimates. The extent to which these biases occur is presumably influenced by the overall proportion of missing values and by whether the probability of missing values depends on further variables, even on θ itself. The model used to estimate θ would have to take these potential dependencies into account.

1.6. Approaches to dealing with missing data in background information

1.6.1. Dummy coding

In several current large-scale assessment programs (Allen et al. [2001]; Foy et al. [2008]; OECD [2009]) missing observations were dummy coded and the dummy variables were subsequently used in the population model. However, Schafer and Graham ([2002]) pointed out that this method merely redefines the parameters as the population model now contains two effects for a variable X in the population model: one effect for the respondents and another effect for the nonrespondents. Moreover, this approach does not take a possible missing at random (MAR) mechanism into account. MAR means that missing values on a certain variable may depend on values of other variables in the model but not on values of the variable itself. Allison ([2002]) demonstrated that even in the case of MCAR, dummy-variable adjustment may lead to biased estimates in ordinary least squares regression when the proportion of missing data is substantial (i.e., 50%).

Rutkowski ([2011]) carried out a simulation study to explore the impact of missing values on background variables on proficiency estimates when missing values on background variables were dummy coded, focusing on the estimation of differences in group means. Under an MAR condition with a maximum of 20% of missing values, the *differences* in group means were estimated without bias. However, both group means were biased in the same direction. Rutkowski termed this a *tandem shift*. Although this tandem shift preserved subgroup differences, it may lead to erroneous conclusions if the results are to be expressed on a predefined scale, for example, to estimate trend effects between different cohorts (Mazzeo & Von Davier [2008]). Rutkowski ([2011]) pointed out that the estimation of subgroup differences can be problematic if the proportions of missing values on the background variables differ between groups. In summary, dummy coding may be appropriate if it is combined with assessment strategies that minimize the proportion of missing values in background information. If this is not the case, defining dummy codes for missing observations may be questionable from a theoretical point of view (Schafer & Graham [2002]) and may also lead to biased mean estimates (Rutkowski, [2011]). Hence, alternative procedures have to be applied.

1.6.2. Single + Multiple Imputation (SMI)

Instead of using dummy codes, missing values on background data may be imputed prior to estimating the parameters of the conditioning model (OECD [2006]; Weirich et al. [2012]. In order to impute the missing data, we have to construct an imputation model which includes all variables that may be related to the occurence of missing data under an MAR assumption. Thus, the population model used to generate the plausible values includes imputed data on the background variables.

The SMI procedure contains a sequence of four steps. The first step is the specification of the item response model (often referred to as calibration). As the aim is to estimate the item parameters only, a simple MML population model without background information can be chosen as $\theta_n = \mu + E_n$, with $E_n \sim N(0, \sigma^2)$. The mean of the population distribution is often defined to be zero, i.e., $\mu = 0$. Hence, to calibrate items, missing values in background variables can be ignored.

The second step is the single imputation of missing values in background variables which requires an imputation model. If we assume that missing values might depend on θ, we should include θ into the imputation model. However, individual θ_n values are estimated in a later step, therefore we have to use a proxy for θ, for example, the percentage of correct responses or point estimates (e.g., MLEs or WLEs) for each person from the item response model. In this study, we will use the latter one.

The third step is the estimation of the conditioning model's parameters. We jointly use the item responses and the (imputed) background information for estimation. In contrast to the item response model, $\theta_n = \mathbf{Y}_\mathbf{N} \mathbf{B} + E_n$. $\mathbf{Y}_\mathbf{N}$ does not contain any missing values because all missing values were imputed in the third step. All background variables are treated as if they were fully observed. A common practice is to consider the item parameters obtained in the first step as the true item parameters (von Davier et al. [2007]).

The fourth and last step is the imputation of plausible values. The imputation model for this step is the conditioning model. Hence, the process of handling different sources of missing data comprises *two* imputation procedures: a single imputation of background variables and a multiple imputation of latent person estimates (θ_n). To avoid confusion between both, we henceforth refer to the first one (step 2) as the imputation model and to the second one (step 4) as the conditioning model. Both imputation steps are not independent because the imputed data from step 2 are used in step 4.

The SMI procedure should help to minimize bias in the population model when missing values on background variables depend on θ. However, there is an uncertainty in the prediction of θ in the population model due to missing values on background information. The SMI procedure may not adequately represent this uncertainty because the population model treats observed and imputed values in the same way. For example, consider two groups with different proportions of missing values on background variables: The uncertainty in the group mean estimates should differ between the groups because we have more observed data in one group than in the other. However, this is not represented in the model if only a single imputation method is used for both groups. Although a bias in estimated group mean differences is unlikely, if the imputation model adequately captures the mechanism behind the missing data, the corresponding standard errors of the mean estimates may be underestimated (White et al. [2010]), in particular if the rates of missing data in background variables are high.

1.6.3. Multiple + Multiple Imputation (MMI)

Nested imputation (Rubin [2003]) or two-stage multiple imputation (Harel [2007]; Harel & Schafer [2003]; Reiter & Drechsler [2007]; Reiter & Raghunathan [2007]) explicitly manages two multiple imputation procedures in a dependent structure. The basic principle is described by Rubin ([2003]):

A few imputations of the first part are created (say *M*), and then for each of these, several imputations of the second part are created (say *N*). The standard combining rules for multiple imputation (the repeated imputation rules) have to be modified because the imputations within a nest are correlated, since they share a common set of imputed values for the first set of missing values. (p. 6)

The nested imputation also allows for different missing value mechanisms, for example, MCAR and MAR (Harel & Schafer [2003]). In the context of large-scale assessments, this may be of particular interest as we may assume different mechanisms in order to account for missing values on θ and missing values on background variables. Furthermore, MMI is preferable to SMI because MMI captures both sources of uncertainty: the uncertainty due to missing values on the background variables and the uncertainty due to the inherently missing θ values.

To implement this method in a large-scale assessment context, we would have to extend the SMI procedure described in the previous section. In the second step, we now create more than one imputation of the background data (which would result in, say, $M = 5$ data sets). Step 3 therefore has to be repeated M times, once for each of the M imputed data sets. In step 4, we would draw N plausible values from each of the M fitted conditioning models, which will result in $M \times N$ plausible values overall. We would then apply the modified combining rules by Rubin ([2003]) to pool statistics based on these plausible values. A more detailed illustration of how the following combining rules are applied to the sets of plausible values is provided in Additional file 1.

Let Q be the quantity of interest (i.e., our set of plausible values). $Q\hat{}(m,n)$

is the mean estimate of the n th plausible value in the m th nest. The overall average $\bar{Q}$,then, is:

$$\bar{Q} = \frac{1}{MN} \sum_{m=1}^{M} \sum_{n=1}^{N} \hat{Q}^{(m,n)}. \tag{3}$$

$\bar{Q}_m$

is a vector of M mean estimates across the N plausible values in each nest

$$\bar{Q}_m = \frac{1}{N} \sum_{n=1}^{N} \hat{Q}^{(m,n)}, \tag{4}$$

and $\bar{U}$

is the overall average of the associated variance estimates,

$$\bar{U} = \frac{1}{MN} \sum_{m=1}^{M} \sum_{n=1}^{N} U^{(m,n)}. \tag{5}$$

Let $MS^{(b)}$ be the between-nest mean square,

$$MS^{(b)} = \frac{N}{M-1}\sum_{m=1}^{M}(\bar{Q}_m - \bar{Q})^2, \tag{6}$$

and let $MS^{(\omega)}$ be the within-nest mean square,

$$MS^{(\omega)} = \frac{1}{M(N-1)}\sum_{m=1}^{M}\sum_{n=1}^{N}\left(\bar{Q}^{(m,n)} - \bar{Q}_m\right)^2. \tag{7}$$

The quantity

$$T = \bar{U} + \frac{1}{N}\left(1 + \frac{1}{M}\right)MS^{(b)} + \left(1 - \frac{1}{N}\right)MS^{(\omega)} \tag{8}$$

estimates the total variance of $(Q - \bar{Q})$

Estimation of the significance levels of Q are based on a Student-t reference distribution, $T^{-0.5}(Q - \bar{Q}) \sim t_{v}$. The degrees of freedom, v, follow from

$$v^{-1} = \left(\frac{N^{-1}(1 + M^{-1})MS^{(b)}}{T}\right)^2 \frac{1}{M-1} + \left(\frac{(1-N^{-1})MS^{(\omega)}}{T}\right)^2 \frac{1}{M(N-1)}. \tag{9}$$

This procedure is computationally more extensive than the commonly applied dummy coding procedure and, to our knowledge, has not been used in any large-scale assessment program yet. However, from a theoretical point of view, MMI strategies seem especially appropriate as they explicitly account for several types of missing data in a dependent structure, which is common in large-scale assessments. The present article therefore assesses whether using MMI leads to more precise estimates of the uncertainty of person parameters than SMI when the proportion of missing values on background variables is substantial. This question is of particular interest if it seems plausible that missing background data are MAR instead of MCAR. The study only investigates the third type of missing data and assumes that missing values in the item responses are MCAR (as a consequence of multiple matrix sampling) and missing values on background variables are MAR. Specifically, our study addressed the following research questions:

1. Does either SMI or MMI of missing values on background variables lead to unbiased estimates of subpopulation proficiency means, the mean subpopulation proficiency difference, and regression coefficients of proficiency on background variables?

2. Does either SMI or MMI of missing values on background variables lead to unbiased estimates of standard errors of subpopulation proficiency means, the mean subpopulation proficiency difference, and regression coefficients?

We expected that both methods would lead to unbiased mean estimates and regression coefficients. Moreover, the MMI method was expected to capture the uncertainty more adequately and was expected to result in slightly higher standard errors and slightly larger confidence intervals for parameters.

2. METHOD

We conducted a simulation study to compare the SMI and the MMI methods when missing proportions on background variables were substantial. Missing values on background variables were assumed to be dependent on θ (i.e., an MAR condition was assumed for the simulation).

2.1 Simulation design

In order to establish realistic conditions for the simulation study, we used item parameters from reading comprehension tasks from the German National Assessment Study 2011 in elementary schools (LV 2011) (Stanat et al. [2012]), a nationally representative German large-scale assessment study. As is common in large-scale studies, the LV 2011 used a balanced multiple matrix sampling design to compile the test booklets. For the present study, 80 dichotomous reading items were grouped into nine disjoint blocks. Each block contained 6 – 16 items (9 items on average). 20 booklets with one or two blocks in each booklet were used in a rotated booklet design to mimic a balanced incomplete block design (see e.g., Frey et al. [2009]; Gonzalez & Rutkowski [2010]). The booklet design is displayed in Table 1. The original item difficulties from the LV 2011 were used to simulate the item responses. The item difficulties ranged from −2.995 to 2.689, which corresponds to a probability for a correct response ranging from 92.9 percent to 9.1 percent. Item responses of $N = 2{,}000$ persons were simulated resulting in 400 valid responses per item.

Table 1. Rotated Block Design To Form Test Booklets

Booklet	Block position 1	Block position 2
1		TL 02
2	TL 03	TL 04
3	TL 01	TL 07
4	TL 05	
5	TL 04	TL 01
6	TL 06	TL 05
7	TL 09	TL 01
8	TL 07	TL 06
9	TL 07	TL 03
10	TL 09	TL 06
11	TL 05	TL 09
12	TL 08	TL 07
13	TL 04	TL 02
14	TL 08	
15	TL 01	TL 03
16		TL 02
17	TL 02	TL 05
18	TL 06	TL 08
19	TL 08	TL 04
20	TL 03	TL 09

Note. TL = testlet. Empty cells mean that the corresponding position in the booklet is non-occupied. For example, booklet 1 only contains testlet 2.
The booklets were used to link the items in the testlets.

The booklets were used to link the items in the testlets.

Two uncorrelated background variables were used in the simulation. The first one, X_1, was uniformly distributed with two levels 0 and 1 (i.e., a grouping variable splitting the population into two subpopulations). The second one, X_2, was normally distributed. The true person proficiency distribution was simulated to depend on both background variables. The population model (see Equation 2) was

$$\theta_n = \beta_1^* X_1 + \beta_2^* X_2 + E_n, \tag{10}$$

with $E_n \sim N(0, \sigma_\theta^2)$ and $\sigma_\theta^2 = 1.44$ Equation 10 is used to both generate and analyze the data. In order to differentiate between the parameters of these models, we use asterisks for the parameters $\beta*1$ and $\beta*2$ of the simulation model to indicate that these are true parameters. In contrast, the parameters of the population model used to analyze the data are denoted β_1 and β_2. For $\beta*1$ and $\beta*2$, a combination of values of 0.1 and 0.4 were used, which resulted in 2 × 2 = 4 conditions for the population model. The coefficients were chosen to mimic conditions which can be expected in real large-scale assessments. In the German National Assessment Study in secondary schools 2012, for example, the largest effect of socioeconomic status (SES) on proficiency is 49 points on the PISA scale with $M = 500$ and $SD = 100$ (Kuhl et al. [2013]). This corresponds to a regression coefficient of about .45. For each background variable, 25% and 40% of the values were deleted, mimicking empirical findings on proportions of missing data for sensitive variables (i.e. SES) when the completion of questionnaires is voluntary (Stanat et al. [2012]). The patterns of missing data of the background variables were in turn created to be dependent on true person proficiencies. For each background variable, the polyserial correlation of the missing pattern with the true person proficiencies was −.10 or −.40. To keep the number of simulations manageable, we refrained from testing against a baseline condition (i.e., 0% missing values or a zero correlation) as we did not expect to find any bias under these conditions.

The simulation was set up using a full-factorial experimental design so that all possible combinations would be represented. We varied the following factors and factor levels:

- Regression coefficients in the population model: $\beta_1^* = \{.1,.4\}$ and $\beta_2^* = \{.1,.4\}$
- Proportions of missing data for both background variables X_1 and X_2: $m(X_1) = \{.25,.4\}$ and $m(X_2) = \{.25,.4\}$
- Correlations between the patterns of missing data in both background variables and θ^*: $d(X_1,\theta^*) = \{-.1,-.4\}$ and $d(X_2,\theta^*) = \{-.1,-.4\}$

In the next step, the 20 booklets were randomly distributed to virtual persons. Using the true person proficiency scores from Equation 10 and the known item parameters, the response probability for person n and item i was defined according to the item response model (see Equation 1):

$$P(X_{ni} = 1) = \frac{e^{\theta_n - b_i}}{1 + e^{\theta_n - b_i}}. \tag{11}$$

Each response X_{ni} was generated by sampling a value from a uniform distribution across the interval [0, 1]. If the sampled value was between 0 and $P(X_{ni} = 1)$, X_{ni} was set to 1. Otherwise, X_{ni} was set to 0.

We used the package TAM (Kiefer et al. [2013]) in R (R Core Team [2014]) to specify a marginal IRT model without conditioning on background variables, where the item parameters were fixed at their known values. This analysis was done for the sole purpose of obtaining a WLE for each person. The WLE served as a rough point estimate of the unknown true proficiency score.

Next, the missing values of both background variables were imputed using the R package MICE (van Buuren & Groothuis-Oudshoorn

[2011]). The imputation model consisted of three variables—X_1, X_2, and the WLE—in a fully conditional specification (see van Buuren [2007]) to account for the process that created the missing data. Two different imputation methods were applied. For X_1, we used a logistic regression imputation according to its binomial distribution. For X_2, we used predictive mean matching. Overall, $M = 5$ imputed data sets were generated. From this point on, the procedure varied depending on whether missing values were imputed under a single or nested imputation method.

2.2 SMI method

The first imputed data set was used to specify the population model in order to estimate the posterior distribution of the person parameters with item parameters fixed at their known values. Using TAM, twenty plausible values were drawn from the posterior distribution. The mean proficiency estimates were pooled across these 20 imputed values according to Rubin ([1987]). The regression estimates were taken from the latent regression part of the population model. Therefore, pooling was not necessary.

2.3 MMI method

The method described above was repeated $M = 5$ times, which resulted in $5 \times 20 = 100$ plausible values. The mean proficiency estimates were pooled over 100 imputed values according to Rubin ([2003]), whereas the regression estimates were pooled over $M = 5$ imputed values according to Rubin ([1987]).

2.4 Measures

The estimates of interest were

- the mean proficiency estimates in both subpopulations (i.e., $\bar{\theta}_{X_1=0}$ and $\bar{\theta}_{X_1=1}$)
- the mean proficiency difference (i.e $\bar{\theta}_{X_1=1}-\bar{\theta}_{X_1=0}$)
- the estimated coefficient β_2 in the latent regression model

The analyses were repeated 1,000 times for each of the 64 conditions. For all estimates mentioned above, three measures were of particular interest in each condition. To examine whether a bias in the corresponding estimates occurred, we computed

$$Bias(x) = N^{-1}\sum_{i=1}^{N}(x_i-x), \tag{12}$$

where N is the number of replications, x_i is the estimate in the i th replication, and x is the true parameter which was used for data generation. For example, if we consider the bias in the regression coefficient β_2, x_i is the estimate of β_2 in the i th replication, and $x = \beta*2$. If we consider the bias in the subpopulation differences, x_i is the estimate of $\bar{\theta}_{X_1=1}-\bar{\theta}_{X_1=0}$ in the i th replication, and x is the true subpopulation difference, i.e. $x = \beta*1$.

The root mean square error (RMSE) between the true x and the estimated x_i parameters were computed by using

$$RMSE(x) = \sqrt{N^{-1}\sum_{i=1}^{N}(x_i - x)^2}. \tag{13}$$

Finally, we were interested in the recovery proportion; that is: How often does the 95% confidence interval include the true value? If the standard error of the estimate had no bias (and the estimate itself had no bias), we would expect that the 95% confidence interval would include the true value in 95% of all cases. The 95% confidence interval for x_i was computed as

$$CI_{95}(x_i) = x_i \pm 1.96 \cdot SE(x_i). \tag{14}$$

To identify whether the bias or the RMSE depended on the simulated conditions, we calculated ANOVA effect sizes across all replications of the simulation. All ANOVA analyses contained the following seven factors, each with two levels: $\beta*1$, $\beta*2$, missing proportion on X_1 ($m(X_1)$), missing proportion on X_2 ($m(X_2)$), dependency between θ^* and the missing values on X_1 ($d(X_1, \theta^*)$), dependency between θ^* and the missing values on X_2 ($d(X_2, \theta^*)$), and the imputation method (SMI vs. MMI). For the sake of conciseness and clarity of results, we only computed main effects, two-way interactions, and three-way interactions in a common model. Hence, $7 + \binom{7}{2} + \binom{7}{3} = 63$ effects were estimated. For the same reason, only effects for which the effect size η^2 exceeded 0.005 are displayed. Moreover, we computed ANOVAs without interaction effects in order to describe whether the addition of interaction effects leads to an increase in explained variance.

To identify whether the recovery proportion depended on the simulated conditions, we conducted a logistic regression analysis. For each of the 1,000 replications of the 64 simulated conditions, we created a dummy code that equaled 1 if the confidence interval of the corresponding estimate incorporated the true value and 0 otherwise. This dummy code was used as the dependent variable in a logistic regression analysis. All regression analyses contain the seven factors mentioned above, each with two levels.

3 RESULTS

The following section summarizes the results of the simulation, focusing on differences between the SMI and the MMI methods. First, we present the bias for the three estimates' mean proficiency difference, β_1 and β_2. We examine whether the bias was influenced by conditions of the simulation using descriptive analyses and ANOVA effect size estimation. The same procedure was carried out for the RMSE and the recovery proportion.

The table for the means of all measures in all of the 64 conditions for both single and nested imputation of background variables is provided in Additional file 1 (Table B1).

3.1 Bias

Tables 2 and 3 display ANOVA effect size estimates of bias in the subpopulation differences (Table 2) and of bias in the regression coefficient β_2 (Table 3). Across all simulated conditions, the bias for all estimates was negligible (mean bias in the subpopulation differences = −0.016; mean bias for regression coefficient β_2 = 0.017). Moreover, the differences between SMI and MMI were negligible (the main effect of imputation method failed to account for a substantial amount of variance in the bias of the subpopulation differences and the regression coefficient β_2). Whereas the overall bias was small, some conditions were associated with higher bias: For example, higher correlations between the patterns of missing data in X_1 and θ^* caused a higher bias in the estimation of subpopulation differences. Higher correlations between the patterns of missing data in X_2 and θ^* caused a higher bias in the estimation of subpopulation differences and in the estimation of β_2. Higher proportions of missing data for background variable X_1 caused a bias in the estimation of β_2. However, most ANOVA effect sizes were small. For example, the seven main effects of the ANOVA analysis displayed in Table 2 together only account for 7.3% of the explained variance, whereas all 64 effects together account for 33.2% of the explained variance. Considering Table 3, the seven main effects only account for 5.7% of the explained variance, whereas all 64 effects account for 20.4% of the explained variance. The bias was therefore rarely associated with a single factor (e.g., a large proportion of missing values alone) but rather with combinations of (adverse) conditions (e.g., a large proportion of missing values *and* a high level of dependence of the pattern of missing data on true person proficiencies). Both SMI and MMI are therefore able to compensate for a single adverse condition. As expected, the imputation method (SMI vs. MMI) did not affect the bias—all ANOVA analyses revealed only negligible main effects of the simulation methods or interaction effects of the simulation method and the other factors.

Table 2. Anova Effect Size Table For Bias In Subpopulation

Differences factor	SS	η^2
$d(X_1, \theta^*) \times d(X_2, \theta^*)$	52.14	0.062
$d(X_1, \theta^*)$	45.18	0.054
$\beta_1^* \times m(X_1)$	21.48	0.025
$\beta_2^* \times d(X_1, \theta^*)$	19.98	0.024
$\beta_1^* \times m(X_1) \times d(X_2, \theta^*)$	17.65	0.021
$m(X_1) \times m(X_2)$	13.54	0.016
$\beta_2^* \times d(X_1, \theta^*) \times d(X_2, \theta^*)$	10.87	0.013
$\beta_2^* \times m(X_1) \times d(X_1, \theta^*)$	7.24	0.009
$\beta_2^* \times m(X_2)$	7.13	0.008
$\beta_2^* \times m(X_2) \times d(X_2, \theta^*)$	6.96	0.008
$\beta_1^* \times \beta_2^* \times d(X_1, \theta^*)$	6.08	0.007
$\beta_1^* \times \beta_2^*$	5.62	0.007
$m(X_1) \times m(X_2) \times d(X_2, \theta^*)$	4.81	0.006
$m(X_1) \times d(X_2, \theta^*)$	4.66	0.006
$d(X_2, \theta^*)$	4.56	0.005
$\beta_2^* \times m(X_1) \times d(X_2, \theta^*)$	4.47	0.005
$m(X_1) \times d(X_1, \theta^*) \times m(X_2)$	4.40	0.005
Residuals	563.55	
Total	844.02	

Table 3. Anova Effect Size Table For Bias In Regression Coefficient B_2

Factor	SS	η^2
$m(X_1)$	12.90	0.087
$\beta_1^* \times \beta_2^*$	4.13	0.028
$d(X_2, \theta^*)$	3.17	0.022
$\beta_1^* \times m(X_1)$	2.42	0.016
β_1^*	2.40	0.016
$\beta_1^* \times \beta_2^* \times d(X_1, \theta^*)$	1.36	0.009
$\beta_1^* \times d(X_2, \theta^*)$	1.19	0.008
$\beta_1^* \times \beta_2^* \times m(X_2)$	1.04	0.007
$\beta_2^* \times m(X_2) \times d(X_2, \theta^*)$	0.82	0.006
Residuals	110.01	
Total	147.56	

3.2 RMSE

Tables 4 and 5 display ANOVA effect size estimates for the RMSE in the subpopulation differences (Table 4) and in the regression coefficient β_2 (Table 5). Across all simulated conditions, the RMSE for the subpopulation differences was .076. The mean RMSE for the regression coefficient β_2 was .035. The RMSE differed between the two imputation methods, but only slightly and only for the regression coefficient β_2: The RMSE was .037 for the single imputation method and .032 for the nested imputation method. This main effect was significant ($\eta^2 = 0.011$, see Table 5). Considering Table 4, the seven main effects of the ANOVA analysis only account for 4.4% of the explained variance, whereas all 64 effects account for 22% of the explained variance. Considering Table 5, the main effects account for 10% of the explained variance, whereas all 64 effects account for 25.3% of the explained variance.

Table 4. Anova Effect Size Table For The Rmse Of Subpopulation Differences

Factor	SS	η^2
$\beta_2^* \times m(X_2) \times d(X_2, \theta^*)$	9.37	0.031
$\beta_1^* \times \beta_2^* \times d(X_1, \theta^*)$	9.31	0.030
$\beta_1^* \times d(X_2, \theta^*)$	6.92	0.023
$m(X_1)$	5.38	0.018
$\beta_1^* \times \beta_2^*$	5.16	0.017
$d(X_1, \theta^*)$	4.58	0.015
$\beta_2^* \times d(X_1, \theta^*) \times m(X_2)$	2.01	0.007
$\beta_1^* \times m(X_1) \times d(X_1, \theta^*)$	1.88	0.006
$\beta_1^* \times d(X_1, \theta^*) \times m(X_2)$	1.85	0.006
$m(X_1) \times m(X_2) \times d(X_2, \theta^*)$	1.62	0.005
Residuals	238.17	
Total	305.29	

Table 5. Anova Effect Size Table For The Rmse Of Regression Coefficient B 2

Factor	SS	η^2
$m(X_1)$	4.34	0.064
β_2^*	1.08	0.016
$\beta_1^* \times \beta_2^*$	0.92	0.013
$d(X_1, \theta^*) \times m(X_2) \times d(X_2, \theta^*)$	0.90	0.013
imp. Method	0.74	0.011
$\beta_1^* \times d(X_1, \theta^*) \times d(X_2, \theta^*)$	0.73	0.011
$m(X_1) \times d(X_1, \theta^*) \times d(X_2, \theta^*)$	0.67	0.010
$\beta_2^* \times d(X_1, \theta^*) \times m(X_2)$	0.65	0.010
$m(X_1) \times m(X_2)$	0.60	0.009
$\beta_1^* \times m(X_1)$	0.56	0.008
$\beta_1^* \times m(X_2)$	0.52	0.008
$\beta_2^* \times m(X_1) \times d(X_1, \theta^*)$	0.52	0.008
$\beta_2^* \times m(X_2)$	0.43	0.006
$\beta_1^* \times \beta_2^* \times d(X_1, \theta^*)$	0.36	0.005
Residuals	50.82	
Total	68.01	

Although we found a significant main effect of imputation method on the RMSE for the regression coefficient β_2, its effect size was small and the reduction of .037–.032 = .005 in the RMSE is negligible.

3.3 Recovery proportion

We conducted logistic regression analyses to predict the probability that the confidence interval of an estimate (e.g., β_2) would include the true value. Tables B2 and B3 are provided in Additional file 1 and display the results of the logistic regression analyses for the recovery proportion of the subpopulation differences (Table B2) and regression coefficient β_2 (Table B3). Comparing the SMI and MMI methods, we found mean recovery proportions of .83 versus .95 for the subpopulation differences, and .85 versus .96 for the regression coefficient β_2. The main effect for the imputation method was significant both for subpopulation differences and for the regression coefficient β_2. MMI overall leads to a higher recovery proportion than SMI. Moreover, the advantage of MMI over SMI is particularly apparent in some specific conditions, for example, if the missing proportion on X_1 is 40% instead of 10% (line 23 in Table B2). This indicates that the MMI method yields a higher recovery proportion if the proportion of missing values is large, or if the correlation between the patterns of missing data in X_2 and θ^* is -.4 instead of -.1. To summarize, MMI seems appropriate for (a combination of) some adverse conditions. This finding reflects the fact that the uncertainty in the imputed values, which results from some adverse conditions, is only incorporated in the MMI method.

4 DISCUSSION

The goal of the present study was to compare two imputation methods—SMI and MMI—for handling missing data in background variables in large-scale assessment programs. The first key question of the present paper was whether these imputation methods would allow unbiased estimation of subpopulation differences and coefficients in the latent regression model. The second question was related to the standard errors of the estimates: How often does the confidence interval include the true value?

Concerning the first research question of unbiased population estimates, none of the simulated conditions showed differences between SMI and MMI. The mean bias was negligible for both imputation methods. Some adverse factors (and especially combinations of adverse factors, e.g., a large proportion of missing values and a high level of dependence of the missing data pattern on true person proficiency) led to biases in the range of −0.145 to 0.092

for the subpopulation differences and −0.041 to 0.061 for the regression coefficient β_2.

Concerning the second research question, when considering the RMSE and especially the recovery proportion, the MMI method outperformed the SMI method. As the MMI method takes the between-imputation variance in the imputed background data into account, the standard errors of the regression coefficient β_2 and of the subpopulation difference accurately reflected the uncertainty due to the amount of missing data. The estimates of the regression coefficient β_2 varied less between the replications and resulted in lower RMSEs. The differences between the two imputation methods, however, were apparent only for substantial proportions of missing data of about 40%, which may occur empirically for some sensitive variables such as SES (Stanat et al. [2012]). Hence, we recommend using MMI instead of SMI if the missing data proportion is substantial *and* if researchers are interested not only in unbiased estimates (e.g., for purely descriptive purposes) but also in unbiased standard errors of these estimates (e.g., for significance testing).

In general, we conclude that imputation methods can provide a suitable alternative to dummy coding, which is flawed by some conceptual problems and may cause biased mean proficiency estimates if the proportion of missing values on the background variables is substantial (Rutkowski [2011]). To date, multiple imputation methods per se are widely established in large-scale assessments to estimate θ_n values, which are considered to be inherently missing in the IRT framework, whereas the problem of missing data in background variables has rarely been tackled by imputation methods. Our analyses show that MMI yields only small practical advantages compared to SMI at the cost of higher computational effort. Therefore, SMI should be a sufficient method in most applications.

Typically, large-scale assessments comprise a multitude of background variables, which are reduced to several uncorrelated principal components prior to the estimation of the population model. Compared to this procedure, the present simulation is highly simplified. In empirical applications where background variables typically are correlated, it is plausible to assume that the more background variables are included, the less problematic is a high proportion of missing values on a single variable as other variables might compensate the loss of information. SMI and MMI methods may easily be combined with principal component analyses, which can be applied after the imputation of missing values in background variables. With MMI, this would result in several data sets, where the number of principal components may vary between the data sets.

A limitation of both SMI and MMI methods is the use of the WLE as a rough proficiency estimate for the unobserved person parameter θ_n in the imputation model. Strictly speaking, the WLE is only an unbiased estimate of the population mean when the same test items are administered to all persons (Wu [2005]). In large-scale assessments, however, this requirement is not fulfilled. Moreover, WLEs suffer from scale unreliability and contain error variance (Mislevy et al. [1992]). Using the WLEs in the imputation model may therefore underestimate the relation between θ and the background variables. This underestimation is expected to be more severe if only few items from the item pool are presented to each person. This would decrease the WLE reliability, which, in turn, leads to a poorer representation of the relation between θ and the background variables in the imputation model. In our simulation, the number of items and persons is constant over all conditions. Consequently, the WLE reliability remained stable at about .70. Future research should investigate the influence of scale reliability when individual θ_n estimates (e.g., WLEs or EAPs) are used in the imputation model to gain information about the level of precision (e.g., in the estimation of subpopulation differences) that could be expected given certain characteristics of the experimental design (e.g., sample size, number of items, number of background variables, or proportion of missing values on background variables). A further improvement of the imputation approach might be to include the unreliability of the WLE as a possible source of uncertainty into the imputation model. However, as a three-level nested imputation this would result in much more computational effort which may not be desirable in empirical applications.

5 CONCLUSION

The present study showed that nested imputation methods are suitable approaches to deal with missing values in both continuous and categorical background variables. Both SMI and MMI yield virtually unbiased estimates of subpopulation differences and regression coefficients for background variables with missing data. Concerning the estimation of standard errors, MMI more accurately reflect the uncertainty due to missing data in the background variables than SMI, resulting in slightly larger standard errors. However, the differences between imputation methods are small in non-extreme conditions of missing data. Thus, both SMI and MMI can be used to impute missing values on background variables in large-scale assessments to avoid the conceptual flaws and the possible biases associated with the common approach of dummy coding missing observations.

ACKNOWLEDGMENTS

This work was supported by the Institute for Educational Quality Improvement at Humboldt-Universität zu Berlin, Berlin, Germany.

The authors would like to thank two anonymous reviewers for their constructive help and support with improving this article.

COMPETING INTERESTS

The authors declare that they have no competing interests.

AUTHORS' CONTRIBUTIONS

SW, NH and MH reviewed the literature. SW, NH, OL and TS designed the analysis. SW conducted the analysis and evaluated the results. SW wrote the manuscript. KB and OL read and revised the manuscript. TS and MH verified the simulation program code. All authors read and approved the final manuscript.

REFERENCES

1. Adams RJ, Wilson M, Wang W-C: **THE MULTIDIMENSIONAL RANDOM COEFFICIENTS MULTINOMIAL LOGIT MODEL.** Applied Psychological Measurement 1997, **21**(1):1–23. 10.1177/0146621697211001
2. Adams RJ, Wu ML: **THE MIXED-COEFFICIENTS MULTINOMIAL LOGIT MODEL: A GENERALIZED FORM OF THE RASCH MODEL.** In Multivariate and Mixture Distribution Rasch Models. Edited by: Davier M, Carstensen CH. Springer, New York; 2007:57–75. 10.1007/978-0-387-49839-3_4

3. Allen NL, Donoghue JR, Schoeps TL: The NAEP 1998 Technical Report. National Center for Educational Statistics, Washington; 2001.

4. Allison PD: Missing Data. Sage Publications, Thousand Oaks, CA; 2002.

5. Bock RD, Aitkin M: **MARGINAL MAXIMUM LIKELIHOOD ESTIMATION OF ITEM PARAMETERS: APPLICATION OF AN EM ALGORITHM.** Psychometrika 1981, **46**(4):443–459. 10.1007/BF02293801

6. De Boeck P: **RANDOM ITEM IRT MODELS.** Psychometrika 2008, **73**(4):533–559. doi: 10.1007/s11336–008–9092-x 10.1007/s11336-008-9092-x

7. Embretson SE, Reise SP: Item Response Theory for Psychologists. Erlbaum, Mahwah, NJ; 2000.

8. Enders CK: Applied Missing Data Analysis. Guilford Press, New York; 2010.

9. Foy P, Galia J, Li I: **SCALING THE DATA FROM THE TIMSS 2007 MATHEMATICS AND SCIENCE ASSSESSMENT.** In TIMSS 2007 Technical Report. Edited by: Olson JF, Martin MO, Mullis IVS. TIMSS & PIRLS International Study Center, Lynch School of Education, Boston College, Chestnut Hill, MA; 2008:225–280.

10. Frey A, Hartig J, Rupp AA: **AN NCME INSTRUCTIONAL MODULE ON BOOKLET DESIGNS IN LARGE-SCALE ASSESSMENTS OF STUDENT ACHIEVEMENT: THEORY AND PRACTICE.** Educational Measurement: Issues and Practice 2009, **28**(3):39–53. 10.1111/j.1745-3992.2009.00154.x

11. Gonzalez E, Rutkowski L: **PRINCIPLES OF MULTIPLE MATRIX BOOKLET DESIGNS AND PARAMETER RECOVERY IN LARGE-**

SCALE ASSESSMENTS. IEA-ETS Research Institute Monograph 2010, **3:** 125–156.

12. Graham JW: **MISSING DATA ANALYSIS: MAKING IT WORK IN THE REAL WORLD.** Annual Review of Psychology 2009, **60:** 549–576. 10.1146/annurev.psych.58.110405.085530

13. Harel O: **INFERENCES ON MISSING INFORMATION UNDER MULTIPLE IMPUTATION AND TWO-STAGE MULTIPLE IMPUTATION.** Statistical Methodology 2007, **4:** 75–89. 10.1016/j.stamet.2006.03.002

14. Harel O, Schafer JL: Multiple Imputation in two Stages. Paper presented at the Proceedings of Federal Committee on Statistical Methodology Research Conference, Washington DC; 2003.

15. Kiefer, T, Robitzsch, A, & Wu, M. (2013). TAM: test analysis modules (R package version 0.5-21). Retrieved from ., [http://CRAN.R-project.org/package=TAM]

16. Kuhl P, Siegle T, Lenski AE: **SOZIALE DISPARITÄTEN [SOCIAL DISPARITIES].** In IQB-Ländervergleich 2012. Mathematische und naturwissenschaftliche Kompetenzen am Ende der Sekundarstufe I. Edited by: Pant HA, Stanat P, Schroeders U, Roppelt A, Siegle T, Pöhlmann C. Waxmann, Münster; 2013. [The IQB National Assessment study 2012—Competencies in mathematics and the sciences at the end of secondary level] (S. 275–296)

17. Little RJA: **REGRESSION WITH MISSING X'S: A REVIEW.** Journal of the American Statistical Association 1992, **87**(420):1227–1237.

18. Lord FM: **ESTIMATION OF LATENT ABILITY AND ITEM PARAMETERS WHEN THERE ARE OMITTED RESPONSES.** Psychometrika 1974, **39**(2):247–264. 10.1007/BF02291471

19. Lord FM: **MAXIMUM LIKELIHOOD ESTIMATION OF ITEM RESPONSE PARAMETERS WHEN SOME RESPONSES ARE OMITTED.** Psychometrika 1983, **48**(3):477–482. 10.1007/BF02293689

20. Ludlow LH, O'leary M: **SCORING OMITTED AND NOT-REACHED ITEMS: PRACTICAL DATA ANALYSIS IMPLICATIONS.** Educational and Psychological Measurement 1999, **59:** 615–630. doi:10.1177/0013164499594004 10.1177/0013164499594004

21. Mazzeo J, Von Davier M: Review of the Programme for International Student Assessment (PISA) Test Design: Recommendations for Fostering Stability in Assessment Results. 2008.

22. Mislevy RJ, Beaton AE, Kaplan B, Sheehan KM: **ESTIMATING POPULATION CHARACTERISTICS FROM SPARSE MATRIX SAMPLES OF ITEM RESPONSES.** Journal of Educational Measurement 1992, **29:** 133–161. 10.1111/j.1745-3984.1992.tb00371.x

23. Assessing Scientific, Reading and Mathematical Literacy. A Framework for PISA 2006. OECD, Paris; 2006.

24. PISA 2006 Technical Report. OECD, Paris; 2009.

25. R: A Language and Environment for Statistical Computing. (Version 3.1.0). R Foundation for Statistical Computing, Vienna, Austria; 2014.

26. Reiter JP, Drechsler J: **RELEASING MULTIPLY-IMPUTED SYNTHETIC DATA GENERATED IN TWO STAGES TO PROTECT CONFIDENTIALITY.** IAB Discussion Paper 2007, **2007**(20):1–18.

27. Reiter JP, Raghunathan TE: **THE MULTIPLE ADAPTIONS OF MULTIPLE IMPUTATION.** Journal of the American Statistical

Association 2007, **102:** 1462–1471. 10.1198/ 016214507 000000932

28. Rubin DB: Multiple Imputation for Nonresponse in Surveys. Wiley, New York; 1987.

29. Rubin DB: **NESTED MULTIPLE IMPUTATION OF NMES VIA PARTIALLY INCOMPATIBLE MCMC.** Statistica Neerlandica 2003, **57**(1):3–18. 10.1111/1467-9574.00217

30. Rutkowski L: **THE IMPACT OF MISSING BACKGROUND DATA ON SUBPOPULATION ESTIMATION.** Journal of Educational Measurement 2011, **48**(3):293–312. 10.1111/j.1745-3984.2011.00144.x

31. Schafer JL, Graham JW: **MISSING DATA: OUR VIEW OF THE STATE OF THE ART.** Psychological Methods 2002, **7:** 147–177. 10.1037/1082-989X.7.2.147

32. Kompetenzen von Schülerinnen und Schülern am Ende der vierten Jahrgangsstufe in den Fächern Deutsch und Mathematik. Waxmann, Münster; 2012.

33. Tuerlinckx F, Rijmen F, Molenberghs G, Verbeke G, Briggs D, Van den Noortgate W, De Boeck P: **ESTIMATION AND SOFTWARE.** In Explanatory Item Response Models. Edited by: Boeck P, Wilson M. Springer, New York; 2004:343–373. 10.1007/978-1-4757-3990-9_12

34. van Buuren S: **MULTIPLE IMPUTATION OF DISCRETE AND CONTINUOUS DATA BY FULLY CONDITIONAL SPECIFICATION.** Statistical Methods in Medical Research 2007, **16:** 219–242. 10.1177/0962280206074463

35. van Buuren S, Groothuis-Oudshoorn K: **MICE: MULTIVARIATE IMPUTATION BY CHAINED EQUATIONS IN R.** Journal of Statistical Software 2011, **45**(3):1–67.

36. von Davier M, Sinharay S, Oranje A, Beaton AE: **THE STATISTICAL PROCEDURES USED IN NATIONAL ASSESSMENT OF EDUCATIONAL PROGRESS: RECENT DEVELOPMENTS AND FUTURE DIRECTIONS.** In Handbook of Statistics: Vol. 26. Psychometrics. Edited by: Rao CR, Sinharay S. Elsevier, Amsterdam; 2007.

37. von Davier M, Gonzalez E, Mislevy RJ: **WHAT ARE PLAUSIBLE VALUES AND WHY ARE THEY USEFUL?** IERI Monograph Series: Issues and Methodologies in Large Scale Assessments 2009, **2:** 9–36.

38. Warm TA: **WEIGHTED LIKELIHOOD ESTIMATION OF ABILITY IN ITEM RESPONSE THEORY.** Psychometrika 1989, **54**(3):427–450. 10.1007/BF02294627

39. Weirich S, Haag N, Roppelt A: **TESTDESIGN UND AUSWERTUNG DES LÄNDERVERGLEICHS: TECHNISCHE GRUNDLAGEN [TEST DESIGN AND ANALYSIS OF IQB NATIONAL ASSESSMENT: TECHNICAL FUNDAMENTALS].** In Kompetenzen von Schülerinnen und Schülern am Ende der vierten Jahrgangsstufe in den Fächern Deutsch und Mathematik. Edited by: Stanat P, Pant HA, Böhme K, Richter D. Waxmann, Münster; 2012. [Students' competences in German and Mathematics at the end of Grade 4: Results of the IQB National Assessment study 2011] (pp. 277–290)

40. White IR, Royston P, Wood AM: **MULTIPLE IMPUTATION USING CHAINED EQUATIONS: ISSUES AND GUIDANCE FOR PRACTICE.** Statistics in Medicine 2010, **30:** 377–399. 10.1002/sim.4067

41. Wilson M, De Boeck P: **DESCRIPTIVE AND EXPLANATORY ITEM RESPONSE MODELS.** In Explanatory Item Response Models. Edited by: Boeck P, Wilson M. Springer, New York; 2004:43–74. 10.1007/978-1-4757-3990-9_2

42. Wu M: **THE ROLE OF PLAUSIBLE VALUES IN LARGE-SCALE SURVEYS.** Studies in Educational Evaluation 2005, **31:** 114–128. 10.1016/j.stueduc.2005.05.005

CHAPTER 3

Spatial Random Slope Multilevel Modeling Using Multivariate Conditional Autoregressive Models: A Case Study of Subjective Travel Satisfaction in Beijing

Guanpeng Dong[a], Jing Ma[b], Richard Harris[c] & Gwilym Pryce[a]

[a] Sheffield Methods Institute, Faculty of Social Sciences, University of Sheffield

[b] School of Geography, Beijing Normal University

[c] School of Geographical Sciences, University of Bristol

ABSTRACT

This article explores how to incorporate a spatial dependence effect into the standard multilevel modeling (MLM). The proposed method is particularly well suited to the analysis of geographically clustered survey data where individuals are nested in geographical areas. Drawing on multivariate conditional autoregressive models, we develop a spatial random slope MLM approach to account for the within-group dependence among individuals in the same area and the spatial dependence between areas simultaneously. Our approach improves on recent methodological advances in the integrated spatial and MLM literature, offering greater flexibility in terms of model specification by allowing regression coefficients to be spatially varied. Bayesian Markov chain Monte Carlo (MCMC) algorithms are derived to implement the proposed model. Using two-level travel satisfaction data in Beijing, we apply the proposed approach as well as the standard nonspatial random slope MLM to investigate subjective travel satisfaction of residents and its determinants. Model comparison results show strong evidence that the proposed method produces a significant improvement against a nonspatial random slope MLM. A fairly large spatial correlation parameter suggests strong spatial dependence in district-level random effects. Moreover, spatial patterns of district-level random effects of locational variables have been identified, with high and low values clustering together.

KEYWORDS

Bayesian estimation, Beijing, conditional autoregressive model, multilevel modeling, subjective travel satisfaction, estimación bayesiana, Beijing, modelo de auto-regresión condicional, modelado a nivel múltiple, satisfacción subjetiva de viaje

Suppose we want to understand what determines perceived travel satisfaction of commuters. We might be interested, for example, in the extent to which travel satisfaction is determined by travel-related variables (commuting time and mode choices) and locational variables (proximity to public amenities) versus the extent to which it is determined by individual life circumstance characteristics such as age, income, and gender. From our data on individuals *i* grouped into areal units *j*, we know that there are likely to be unobserved similarities and connections between individuals in the same area (Browne and Goldstein 2010). A modeling framework that explicitly acknowledges the multilevel structure of the data in terms of lower (individual *i*) and higher levels (areal unit *j*) will be required. An obvious solution would be to adopt a standard multilevel modelling (MLM) framework of the kind proposed by Goldstein (2003). MLM not only naturally captures the dependence effect within each group but offers a flexible framework for modeling heterogeneity of covariate effects whereby the slopes of covariates are allowed to vary across groups.

There is a potential problem, however, when applying MLM to hierarchical data where the study area is made up of groups defined by geographical areas such as districts or regions. In this context, although the vertical or hierarchical group dependence is modeled in an MLM, the horizontal dependence among areas is not (Dong and Harris 2015). More specifically, random effects including regression intercepts and slopes of one area are assumed to be uncorrelated with those of areas nearby even when they are geographically adjacent. That is, MLM conceptualizes geography simply as "place" through which group membership structure is defined, ignoring the dimension of "space"—the interplay and interactions between areas and the people who live in them (Arcaya et al. 2012; Owen, Harris, and Jones 2015).

Some arguments for the inclusion of spatial dependence in the area-level random effect have been put forward. For example, Morenoff (2003) and Savitz and Raudenbush (2009) argued that individuals' outcomes in one area could be affected by what happens in nearby areas because boundaries of spatial units might not be strictly impermeable. In other words, individuals might be living in different areas but in close proximity if they are located on either side of a geographical boundary and so might experience broadly the same events in the local area (Haining 2003). Second, geographical outcomes might be varying continuously across areas because they simply do not follow geographical boundaries (Fotheringham, Brunsdon, and Charlton 2002; Haining 2003). Third, the geographical boundaries used to designate areas at the various levels of a multilevel model are often arbitrary or determined by data availability. If, as is likely, there is a mismatch between the chosen spatial units and the spatial scales at which certain social processes under exploration operate, spatial dependence tends to occur (Anselin 1988; Fotheringham, Brunsdon, and Charlton 2002). Failing to account for spatial dependence will lead to unreliable estimates for both variance parameters and regression coefficients in MLM (Dong et al. 2015).

With increasing awareness of the problems associated with applying MLM to geographically hierarchical data, there is emerging interest in the development of hybrid methodologies to incorporate spatial dependence effect into MLM (Owen, Harris, and Jones 2015). Notably, in the spatial econometrics literature, efforts have recently been made to develop integrated spatial econometric and MLM frameworks (Corrado and Fingleton 2012; Baltagi, Fingleton, and Pirotte 2014; Dong and Harris 2015). Regarding these methodological advances, spatial dependence is conceptualized as a simultaneous autoregressive (SAR) model following the spatial econometrics convention (e.g., Cliff and Ord 1981; Anselin 1988; LeSage and Pace 2009). In another type of hybrid model, spatial dependence (at the lower level) is represented by a conditional autoregressive (CAR; e.g., Besag, York, and Mollie 1991; Banerjee, Carlin, and Gelfand 2004) model in addition to another set of independent random effects at the higher level (Arcaya et al. 2012). Similarly, Browne, Goldstein, and Rasbash (2001) employed a spatial multiple membership model to tackle spatial dependence effects in the sense that individuals are allowed to be influenced by both their immediate and neighboring contexts. Other conceptualizations of spatial dependence such as spatial Gaussian processes (e.g., Cressie 1993; Haining 2003; Banerjee, Carlin, and Gelfand 2004) have been included in MLM by Chaix, Merlo, and Chauvin (2005) and Chaix, Merlo, Subramanian, et al. (2005). A brief summary on differences between these methodologies is provided in Dong et al. (2015).

The baseline model on which the aforementioned extensions build is the standard random intercept MLM, which means that neither varying covariate effects nor spatial dependence in covariate effects is considered. This represents a serious limitation. Research questions of how covariate effects are heterogeneous across areas might be of great interest to researchers and policymakers. Although the standard random slope MLM is suitable for this type of research with multilevel data structures if higher level units are independent of each other (Raudenbush and Bryk 2002; Goldstein 2003), how to incorporate spatial dependence into the standard random slope MLM for investigating spatially varying covariate effects has rarely been studied.

This article fills the gap by proposing a spatial random slope MLM in which the higher level random effects are conceptualized as a multivariate CAR (MCAR) model. With a complex MCAR model (e.g., Besag, York, and Mollie 1991; Banerjee, Carlin, and Gelfand 2004), both the spatial dependence across areas and the within-area correlations between different sets of random covariate effects can be simultaneously accommodated. There are several important advantages associated with the proposed model. Generally, it will provide more efficient and accurate estimation for regression coefficients when compared to the standard nonspatial random slope MLM. In addition, MCAR models can introduce spatial smoothing and the borrowing of strength across neighboring areal units, leading to estimates of random covariate effects that are robust and have higher precision, which is of particular importance for areas with small sample sizes (Neelon, Anthopolos, and Miranda

2014). Furthermore, as correlations between different sets of random covariate effects are considered in the developed approach, the derivation of one set of random effects can benefit from the estimation of other sets of random effects (Congdon 2014). Finally, we contribute to the ongoing integrated spatial and MLM literature by proposing a flexible approach that simultaneously considers spatially varying intercepts and covariate effects.

In a simpler single-level modeling setting, MCAR models have been employed to examine spatially heterogeneous covariate effects across a study region in the so-called spatially varying coefficient models (e.g., Assuncao 2003; Congdon 2014). A close counterpart of our model is the one developed by Gelfand et al. (2007), which is motivated by a very specific set of house price data where residential apartments nest into spatially referenced building projects. Their higher level units are represented by spatial point objects and a multivariate spatial Gaussian process is employed to model the project-level random effects. In contrast, the higher level units in our case study are areal units (census districts in Beijing, China), as in many other geographically clustered survey data, and random effects at the area level are represented by the MCAR model, which is more suitable than the spatial Gaussian processes (Banerjee, Carlin, and Gelfand 2004; Wall 2004).

Using Beijing as a case study, we use the proposed method to investigate spatial variations of individuals' subjective travel satisfaction and its determinants. The data are derived from a large satisfaction survey of Beijing in 2005, which has a two-level structure with individuals nested into districts. We first comprehensively investigate how sociodemographic attributes, travel-related characteristics, and locational variables influence individuals' subjective travel satisfaction. Our main interest, however, is in whether and how the effects of locational factors (e.g., proximity to green parks and subway accessibility) vary across districts. The case study shows strong evidence that the proposed model provides a significant improvement against the nonspatial random slope MLM. Moreover, it also presents the spatial patterns of district-level random effects of locational variables with high and low values clustering together, which has rarely been researched before.

The remainder of the article is organized as follows. First, we introduce the technical foundations for our model, including the basics of the standard random slope MLM, followed by the CAR and MCAR models. We then specify the spatial random slope MLM and derive Bayesian Markov chain Monte Carlo (MCMC) algorithms for model implementation. Next, we apply the proposed approach to investigate the determinants of subjective travel satisfaction in Beijing, China. Finally, we conclude with a brief summary of research findings and suggestions for future work.

1. MODEL FOUNDATIONS: RANDOM SLOPE MLM, CAR, AND MCAR

1.1 The Standard Nonspatial Random Slope MLM

Consider two-level clustered survey data where individuals nest into areas, at some more aggregate scale. Following Goldstein (2003), the standard nonspatial random slope MLM is specified as

$$y_{ij} = \boldsymbol{\beta}_{0j} + \boldsymbol{\beta}_{1j} x_{ij} + \varepsilon_{ij} \qquad (1)$$

$$\boldsymbol{\beta}_{0j} = \boldsymbol{\beta}_0 + x_j \gamma_0 + u_{0j} \qquad (2)$$

$$\beta_{1j} = \beta_1 + x_j \gamma_1 + u_{1j} \tag{3}$$

$$\operatorname{var}(\varepsilon_{ij}) = \sigma_e^2;\ \operatorname{cov}(u_{0j}, u_{1j}) = V_u = \begin{bmatrix} \sigma_{u0}^2 & \sigma_{u01}^2 \\ \sigma_{u10}^2 & \sigma_{u1}^2 \end{bmatrix}, \tag{4}$$

where i and j indicate individual and higher levels, respectively; y_{ij} is the outcome value for the ith individual within the jth higher level unit; x_{ij} is an individual-level covariate and β_{1j} are the associated regression slopes that vary across higher level units; β_{0j} are the random intercepts; and ϵ_{ij} is the individual-level residuals that are assumed to follow an independent normal distribution, $N(0, \sigma_e^2)$.

Equations 2 and 3 relate the random effects to a higher level covariate (x_j). In other words, higher level covariates can be included to explain the heterogeneity in the two sets of random effects (β_{0j} and β_{1j}). The term γ_0 measures the effect of x_j in explaining the random intercepts, and γ_1 quantifies the effect of x_j in explaining the random slopes of x_{ij}. The unexplained heterogeneity is denoted by u_{0j} and u_{1j}, respectively. The parameters σ_{u0}^2 and σ_{u1}^2 measure the amount of variation in the intercept and regression slope, and σ_{u01}^2 is the covariance of these two sets of random effects.

The resultant marginal distribution for u_{0j} is an independent normal distribution $N(0, \sigma_{u0}^2)$ and that for u_{1j} is $N(0, \sigma_{u1}^2)$. The covariance parameter σ_{u01}^2 is interpreted as the within-area association between u_{0j} and u_{1j} with the correlation coefficient measured by $\sigma_{u10}^2/\text{sqrt}(\sigma_{u0}^2 \times \sigma_{u1}^2)$. As previously mentioned, when the groups are areal units, we also anticipate the existence of between-area associations in each set of higher level random effects, which occur because of the geographical proximity-based spatial dependence effect.

1.2 CAR Models

Following Banerjee, Carlin, and Gelfand (2004), a J by J spatial weights matrix or neighborhood structure for higher level spatial units, $W = (w_{jk})$ is usually defined by geographical contiguity: $w_{jk} = 1$ if the jth and the kth areas share boundaries (denoted by $j \sim k$) and 0 otherwise. The intrinsic CAR (ICAR) is defined by the following full conditional distributions (Besag, York, and Mollie 1991; Congdon 2014):

$$\mathrm{E}\left(u_j \mid \boldsymbol{u}_{-j}\right) = \frac{1}{w_{j+}} \sum\nolimits_{k \sim j} u_k,$$

$$\operatorname{prec}\left(u_j \mid \boldsymbol{u}_{-j}\right) = \tau^2 w_{j+},\ j = 1, \ldots, J, \tag{5}$$

where w_{j+} is the number of neighbors of the jth area; $u_j = (u_1, \ldots, u_{j-1}, u_{j+1}, \ldots, u_J)$ indicates random effects other than the jth area; the scalar τ^2 is the conditional precision parameter and so the variance of random effects $\boldsymbol{u}$ is $1/\tau^2$; the conditional expectation of u_j, $E(u_j \mid \boldsymbol{u}_{-j})$ is the mean of random effects from surrounding areas; and the conditional precision $\operatorname{prec}(u_j \mid \boldsymbol{u}_{-j})$ is proportional to the number of neighbors w_{j+}.

It is well known that the full conditional distributions for each observation together give rise to a unique intrinsic Gaussian Markov random field (GMRF), $\boldsymbol{u} \sim MVN\ (\mathbf{0}, \Omega_{\mathrm{ICAR}})$ (Besag, York, and Mollie 1991; Rue and Held 2005). The J by J precision matrix Ω_{ICAR} is

$$\Omega_{ICAR} = \tau^2 (D_W - W), \tag{6}$$

where D_W = diag (w_{1+}, w_{2+}, . . . , w_{J+}). Therefore, the density distribution of $\boldsymbol{u}$ is given by

$$p(\boldsymbol{u}) = (2\pi)^{-J/2}(|\Omega_{\text{ICAR}}|^{*})^{1/2}\exp(-1/2\boldsymbol{u}'\Omega_{\text{ICAR}}\boldsymbol{u}). \tag{7}$$

The precision matrix Ω_{ICAR} is improper because the row sums of (D_W - W) are equal to a vector of zeros. $|\Omega_{\text{ICAR}}|^*$ denotes the generalized determinant calculated by the product of the (n - 1) nonzero eigenvalues of Ω_{ICAR} (Rue and Held 2005). In implementation, a sum-to-zero constraint on random effects should be enforced in each of the MCMC iterations (Banerjee, Carlin, and Gelfand 2004).

As an alternative, Leroux, Lei, and Breslow (1999) proposed a new CAR formulation (LCAR), specified as

$$\begin{aligned} E(u_j \mid \boldsymbol{u}_{-j}) &= \frac{\lambda}{1-\lambda+\lambda w_{j+}}\sum_{k\sim j} u_k,\ \text{prec}(u_j \mid \boldsymbol{u}_{-j}) \\ &= \tau^2(1-\lambda+\lambda w_{j+}),\ 0 \le \lambda \le 1. \end{aligned} \tag{8}$$

The resultant precision matrix for the LCAR specification is $\Omega_{\text{LCAR}} = \tau^2(L_W - \lambda W)$, where L_W = diag (1 - λ + λw_{j+}). The LCAR model corresponds to an ICAR model when λ equals 1, reducing to an independent normal distribution when λ equals 0. At its heart, LCAR conceptualizes two sources of variability in area effects, the first being spatially structured and characterized by an ICAR process and the other being randomly distributed, characterized by an independent normal distribution (MacNab 2011). The parameter λ serves as a weight parameter measuring the relative importance between the two sources and as an indicator of the intensity of spatial dependence (Congdon 2014).

Other types of GMRF models such as proper CAR and convolution CAR (or the BYM model; Besag, York, and Mollie 1991) have also been extensively used to model spatial dependence (for a comprehensive technical review, see Banerjee, Carlin, and Gelfand [2004] and Congdon [2014]). One of the appealing theoretical properties of LCAR is that it can properly represent various levels of spatial dependence with a single set of random effects (Lee 2011). In contrast, ICAR models are only appropriate in situations when spatial dependence is very strong, which is self-evident as the spatial correlation parameter is set to one. The BYM model, on the other hand, uses the sum of two independent sets of random effects (one being ICAR and the other being independent normal) to represent spatial dependence. This complex conceptualization of spatial dependence makes BYM suffer from a lack of model identifiability, normally leading to large uncertainty in estimated model parameters (MacNab 2011). Technically, simulation studies have found that the LCAR model outperforms other CAR models including ICAR, proper CAR, and BYM in terms of both retrieving predefined spatial parameters and covariate effects across a wide range of spatial dependence scenarios (Lee 2011). Similar arguments and suggestions of using LCAR and its multivariate version to model spatial dependence have been made in MacNab (2011). As such, this article focuses on the LCAR model and its multivariate version for incorporating spatial dependence into the standard nonspatial random slope MLM. For the purpose of an easy exposition of complex MCAR models, ICAR and its multivariate version are also discussed in the article.

1.3 MCAR Models

In this section, multivariate ICAR (MICAR) and multivariate LCAR (MLCAR) models are briefly reviewed. For a general discussion on MCAR, see Gelfand and Vounatsou (2003), Banerjee, Carlin, and Gelfand (2004), Martinez-Beneito (2013), and Congdon (2014), among others.

Suppose there are P random effects for each area j (here we have P random covariate slopes) and denote $\boldsymbol{u} = (\boldsymbol{u}_1, \boldsymbol{u}_2, \ldots, \boldsymbol{u}_J)$ with $\boldsymbol{u}_j = (u_{j1}, u_{j2}, \ldots, u_{jP})$, $j = 1, 2, \ldots, J$, $p = 1, 2, \ldots, P$. Following Congdon (2014), the conditional distribution for $\boldsymbol{u}_j$ in MICAR is

$$E(\boldsymbol{u}_j|\boldsymbol{u}_{-j}) = \frac{1}{w_{j+}}\sum_{k\neq j} w_{jk}\mathbf{I}_P\boldsymbol{u}_k,\ \text{prec}(\boldsymbol{u}_j|\boldsymbol{u}_{-j}) = w_{j+}\Gamma, \tag{9}$$

where $\boldsymbol{I}_P$ is a P-dimensional identity matrix, Γ is a P by P positive definite precision matrix, and its inverse, Σ, is the familiar covariance matrix. The diagonal entries of Σ are variances of each set of random effects and the off-diagonal entries measure the within-area correlations between different sets of random effects. As in ICAR, full conditionals specified in Equation 9 together result in a unique intrinsic multivariate GMRF, $\boldsymbol{u} \sim MVN\ (\mathbf{0}, \Omega_{\text{MICAR}})$ with a JP by JP precision matrix defined as

$$\Omega_{\text{MICAR}} = (D_W - W) \otimes \Gamma, \tag{10}$$

where the symbol ⊗ denotes the Kronecker product (Banerjee, Carlin, and Gelfand 2004). Similarly, the conditional distribution of $\boldsymbol{u}_j$ in MLCAR is (Congdon 2014)

$$\mathrm{E}(\boldsymbol{u}_j \mid \boldsymbol{u}_{-j}) = \frac{\lambda}{1-\lambda+\lambda w_{j+}} \sum_{k \neq j} w_{jk} I_P \boldsymbol{u}_k, \ \mathrm{prec}(\boldsymbol{u}_j \mid \boldsymbol{u}_{-j}) = (1-\lambda+\lambda w_{j+})\Gamma. \quad (11)$$

The precision matrix for the MLCAR is $\Omega_{\mathrm{MLCAR}} = (L_W - \lambda W) \otimes \Gamma$. In contrast, the precision matrix for the nonspatial random slope MLM is simply $\boldsymbol{I}_n \otimes V_u^{-1}$, where $\boldsymbol{I}_n$ is an identity matrix of order n. In the subsequent analyses, MLCAR will be incorporated into nonspatial random slope MLM to model the spatial dependence among higher level random effects.

2. PROPOSING A SPATIAL RANDOM SLOPE MLM APPROACH

2.1 Model Specification

Proceeding with the notation in the standard nonspatial random slope MLM, a spatial random slope MLM is specified as

$$\begin{aligned}
y_{ij} &= \boldsymbol{\beta}_{0j} + \boldsymbol{\beta}_{1j} x_{ij} + \varepsilon_{ij} \\
\boldsymbol{\beta}_{0j} &= \boldsymbol{\beta}_0 + x_j^T \gamma_0 + u_{0j} \\
\boldsymbol{\beta}_{1j} &= \boldsymbol{\beta}_1 + x_j^T \gamma_1 + u_{1j} \\
\mathbf{E}\left[\boldsymbol{u}_j = \begin{pmatrix} u_{0j} \\ u_{1j} \end{pmatrix} \middle| \boldsymbol{u}_{-j}\right] &= \frac{\lambda}{1-\lambda+\lambda w_{j+}} \sum_{k \neq j} w_{jk} I_2 \boldsymbol{u}_k. \quad (12)
\end{aligned}$$

Note that the MLCAR model for the two sets of area-level random effects $\boldsymbol{u}_0$ and $\boldsymbol{u}_1$ is used in Equation 12 given the MICAR model is a just special case of the MLCAR with λ equal to one. For the purpose of the simplicity, a spatial random slope MLM with the MLCAR model is labeled as MLM-MLCAR, and a spatial random slope MLM with the MICAR model is labeled as MLM-MICAR. Under MLM-MLCAR, arranging the two sets of random effects by areas, as in Equation 9, the distribution of random effects $\boldsymbol{u}$ is MVN $(0, \Omega_{\text{MLCAR}})$, where $\Omega_{\text{MLCAR}} = (L_W - \lambda W) \otimes \Gamma$. The covariance of $\boldsymbol{u}$ is $\Omega_{\text{MLCAR}}^{-1} = [(L_W - \lambda W) \otimes \Gamma]^{-1} = [(L_W - \lambda W)]^{-1} \otimes \Gamma^{-1}$. The inverse of the precision matrix Γ, Γ^{-1}, is similar to V_u in the nonspatial random slope MLM in Equation 4. Spatial dependence effects in regression slopes are not modeled because of the missing spatial component $[(L_W - \lambda W)]^{-1}$ in the covariance matrix from the nonspatial random slope MLM. It is useful to note that higher level residuals are assumed to be uncorrelated with residuals at the individual level in the proposed model to avoid the problem of identifiability.

2.2 Model Implementation

The proposed MLM-MLCAR is implemented within the Bayesian framework, using a Gibbs sampler with Metropolis updates when required (Banerjee, Carlin, and Gelfand 2004; Gelman et al. 2004). The choice of a Bayesian approach over a frequentist one for model implementation lies in that the model fits nicely into the Bayesian hierarchical spatial modeling framework (Banerjee, Carlin, and Gelfand 2004). Although frequentist approaches are possible for the proposed model, Bayesian estimation enables exact inference on model parameters and appropriate model uncertainty assessment (Gelfand 2012). To derive the full conditional posterior distributions of each model parameter, we rewrite the model in a compact matrix format:

$$y = X\boldsymbol{\beta} + Z\boldsymbol{\theta} + \varepsilon; \ \boldsymbol{\theta} = [\boldsymbol{\theta}_1, \ \boldsymbol{\theta}_2, \ \ldots, \ \boldsymbol{\theta}_J], \tag{13}$$

where y is an N by 1 outcome vector, X is a covariate design matrix, which can consist of lower covariates, higher level covariates, and potential interaction terms between them. For example, if there are p low-level covariates and q higher level ones, then X will be an N by pq matrix. The pq by 1 vector $\boldsymbol{\beta}$ are fixed regression coefficients to estimate. Z is an N by Jp random effect design matrix, which is a block diagonal matrix with each block containing the lower level covariate values of individuals within one area. The Jp by 1 vector $\boldsymbol{\theta}$ stores the random effects arranged by area.

2.3 Prior Distributions

For Bayesian inference, the basic principle is that the posterior distribution of $\boldsymbol{\theta}^*$ (unknown model parameters) is proportional to the product of the data likelihood and prior distributions (Gelman et al. 2004), as denoted by

$$\mathrm{P}\,(\boldsymbol{\theta}^* \,|\, (\mathrm{Data}) \propto \mathrm{P}\,(\mathrm{Data} \,|\, \boldsymbol{\theta}^*) \times \mathrm{P}\,(\boldsymbol{\theta}^*). \tag{14}$$

For the proposed approach here, the unknown model parameter vector is $\boldsymbol{\theta}^* = \{\boldsymbol{\beta}, \boldsymbol{\theta}, \lambda, \sigma_e^2, \Gamma\}$. For prior distributions, a k-dimensional multivariate normal prior is set for $\boldsymbol{\beta}$, with mean M_0 and covariance T_0. Therefore, $P\,(\boldsymbol{\beta}) \sim MVN(\mathrm{M}_0, \mathrm{T}_0)$. The individual-level variance σ_e^2 is assigned an inverse gamma distribution, $P\,(\sigma_e^2) \sim IG\,(c_0, d_0)$, and the area-level precision matrix Γ is assigned a Wishart distribution, $P\,(\Gamma) \sim dwish\,(R_0, v_0)$. These priors are conjugate priors so that the posterior distributions of parameters are in the same family as the prior probability distributions and are employed mainly for computational convenience. A uniform prior distribution is assigned for λ over (0, 1). Following Gelman et al. (2004), the inverse gamma distribution with c_0 and d_0 being the shape parameter and scale parameter is parameterized as

$$P(\sigma_e^2) \propto \sigma_e^{2-c_0-1} \exp(-d_0/\sigma_e^2), \tag{15}$$

and a Wishart distribution with a scale matrix R_0 and a degree of freedom parameter v_0 is specified as

$$P(\Gamma) \propto |\Gamma|^{(v_0-p-1)/2} \exp\{-1/2\mathrm{tr}(R_0{}^{-1}\Gamma)\}. \tag{16}$$

The likelihood function for a spatial random slope MLM is

$$L(Y \mid \boldsymbol{\beta}, \boldsymbol{\theta}, \lambda, \sigma_e^2, \Gamma) = (2\pi\sigma_e^2)^{-N/2} \exp\left\{-(2\sigma_e^2)^{-1}(Y - X\boldsymbol{\beta} - Z\boldsymbol{\theta})'(Y - X\boldsymbol{\beta} - Z\boldsymbol{\theta})\right\}. \tag{17}$$

Therefore, the full posterior distribution for $\boldsymbol{\theta}^* = \{\boldsymbol{\beta}, \boldsymbol{\theta}, \lambda, \sigma_e^2, \Gamma\}$ is

$$P(\boldsymbol{\beta}, \boldsymbol{\theta}, \lambda, \sigma_e^2, \Gamma \mid Y)\ L(Y \mid \boldsymbol{\beta}, \boldsymbol{\theta}, \lambda, \sigma_e^2, \Gamma) \times P(\boldsymbol{\beta}) \times P(\lambda) \times P(\boldsymbol{\theta} \mid \lambda, \Gamma) \times P(\Gamma) \times P(\sigma_e^2). \tag{18}$$

2.4 The Full Conditional Posterior Distribution of β

Let the posterior distribution for **β** be P (**β** | Y, ***θ***, λ, σ_e^2, Γ) ~ *MVN* (M_β, $\sum_\beta$). According to Equation 18, the posterior distribution for **β** is

$$P(\boldsymbol{\beta} \mid Y, \boldsymbol{\theta}, \lambda, \sigma_e^2, \Gamma) \propto L(Y \mid \boldsymbol{\beta}, \boldsymbol{\theta}, \lambda, \sigma_e^2, \Gamma) \times P(\boldsymbol{\beta})$$
$$\propto \exp\left\{-\left(2\sigma_e^2\right)^{-1}(Y - X\boldsymbol{\beta} - Z\boldsymbol{\theta})'(Y - X\boldsymbol{\beta} - Z\boldsymbol{\theta})\right\} \times \exp\left\{-\frac{1}{2}(\boldsymbol{\beta} - M_0)'T_0^{-1}(\boldsymbol{\beta} - M_0)\right\}$$
$$\propto \exp\left\{-\frac{1}{2} \times \boldsymbol{\beta}'\left[\left(\sigma_e^2\right)^{-1}X'X + T_0^{-1}\right]\boldsymbol{\beta} + \left[\left(\sigma_e^2\right)^{-1}(Y - Z\boldsymbol{\theta})'X + T_0^{-1}M_0\right]\boldsymbol{\beta} + con\right\},$$

(19)

where *con* is a constant value regardless of **β**. Comparing this with a standard multivariate normal density, we obtain

$$\sum\nolimits_{\boldsymbol{\beta}} = \left[\left(\sigma_e^2\right)^{-1}X'X + T_0^{-1}\right]^{-1};$$
$$M_{\boldsymbol{\beta}} = \sum\nolimits_{\boldsymbol{\beta}}\left[\left(\sigma_e^2\right)^{-1}X'(Y - Z\boldsymbol{\theta}) + T_0^{-1}M_0\right].$$

(20)

2.5 The Conditional Posterior Distributions for {θ, σ_e^2, Γ}

Following procedures similar to those used in deriving the full posterior distribution for ***β***, we show the full posterior distribution for ***θ*** is also *MVN* (M_θ, $\sum_\theta$), where

$$\sum\nolimits_{\theta} = \left[(\sigma_e^2)^{-1} Z'Z + (L_W - \lambda W) \otimes \Gamma \right]^{-1};$$
$$M_\theta = \sum\nolimits_{\theta} \left[(\sigma_e^2)^{-1} Z'(Y - X\boldsymbol{\beta}) \right]. \tag{21}$$

The full conditions for σ_e^2 is $IG\ (c_e, d_e)$, where

$$c_e = N/2 + c_0; d_e = \frac{1}{2} \times (Y - X\boldsymbol{\beta} - Z\boldsymbol{\theta})' \times (Y - X\boldsymbol{\beta} - Z\boldsymbol{\theta}) + d_0. \tag{22}$$

The full conditional distribution for the within-area precision matrix Γ is *dwish* (R^*, v^*) where

$$v^* = J + v_0; R^* = \left[R_0^{-1} + \boldsymbol{\theta}^{*'} (L_W - \lambda W) \boldsymbol{\theta}^* \right]^{-1} \tag{23}$$

in which $\boldsymbol{\theta}^*$ is a J by p matrix with each column being the random effects that pertain to each individual-level covariate.

2.6 The Conditional Posterior Distribution for λ

The posterior condition distribution for λ, $P\ (\lambda \mid Y, \boldsymbol{\beta}, \boldsymbol{\theta}, \sigma_e^2, \Gamma) \propto P\ (\boldsymbol{\theta} \mid \lambda, \Gamma) \times P\ (\lambda)$ is

$$\propto |(L_W - \lambda W) \otimes \Gamma|^{1/2} \exp\left\{ -1/2\boldsymbol{\theta}'[(L_W - \lambda W) \otimes \Gamma]\boldsymbol{\theta} \right\}. \quad (24)$$

Unlike $\{\boldsymbol{\beta}, \boldsymbol{\theta}, \sigma_e^2, \Gamma\}$, the spatial correlation parameter λ does not have a standard density distribution from which samples can be easily drawn. Therefore, an adaptive Metropolis approach with an acceptance rate of about 50 percent is used to update λ (Chib and Greenberg 1995; Gelman et al. 2004).

3. APPLICATION OF MLM-MLCAR TO SUBJECTIVE TRAVEL SATISFACTION

In this section, the proposed MLM-MLCAR is employed to examine social and spatial variations in individuals' subjective travel satisfaction in urban Beijing, China. Based on a large-scale satisfaction survey conducted in 2005, we comprehensively investigate the impacts of sociodemographic attributes, travel-related characteristics, and locational variables on individuals' subjective travel satisfaction. Furthermore, we seek to understand how geographical context might influence covariate effects on travel satisfaction, which is an underresearched area.

3.1 Data and Variables

Our data are derived from a large-scale household satisfaction survey conducted in Beijing that collects residents' sociodemographics, travel-related characteristics, and evaluation of general living environment. The purpose was to evaluate Beijing's general livability, including life convenience, travel satisfaction, human and physical environment suitability, and health and safety conditions. The target population were residents living in urban Beijing (including 134 districts or *Jiedao* in total) for at least six months who are presumed to be familiar enough with their living environment to provide a good assessment. The survey had a stratified random sampling process, with the sample size in each district approximately proportional to its total population. In total, 11,000 questionnaires were issued and 7,647 of them were returned, of which 6,544 were valid. Further details of the survey and sample profiles are provided in Zhang, Yin, and Zhang (2006). It has been reported that this survey is quite representative of the overall characteristics of Beijing's population, when compared to the population census data (Zhang, Yin, and Zhang 2006; Meng, Zheng, and Yu 2011).

In this study, the overall travel satisfaction of each individual is modeled as the outcome of interest. Daily travel satisfaction is an important component of emotional well-being, and it might have a significant influence on overall life satisfaction (Ettema et al. 2012). Moreover, subjective travel satisfaction seems to be a more comprehensive and welfare-related indicator of effectiveness of urban planning and transport policy, when compared to objective evaluation measures such as average commuting time or traffic congestion (Meng, Zheng, and Yu 2011). As Beijing is experiencing rapid urban expansion, growing car use, increasing transport carbon emission, and serious air pollution, a better investigation and understanding of what factors influence subjective travel satisfaction and how geographical context shapes its spatial variation is of great value to urban planning, transport development, and relevant public policy evaluation in urban China (Ma et al. 2014).

Table 1 presents the specific survey questions on travel satisfaction, including how people rate their satisfaction with public transport convenience, commuting convenience, and non-work-related travel convenience. For each of the survey questions, residents' responses are measured on a five-category Likert scale ranging from *very unsatisfied* to *very satisfied.* The overall travel satisfaction of each individual, defined as the outcome variable in the model, is calculated by averaging scores on each specific survey question listed in Table 1. The mean travel satisfaction scores approximate well to a continuous normal distribution (Figure 1) and thus are modeled as a continuous variable in this study.

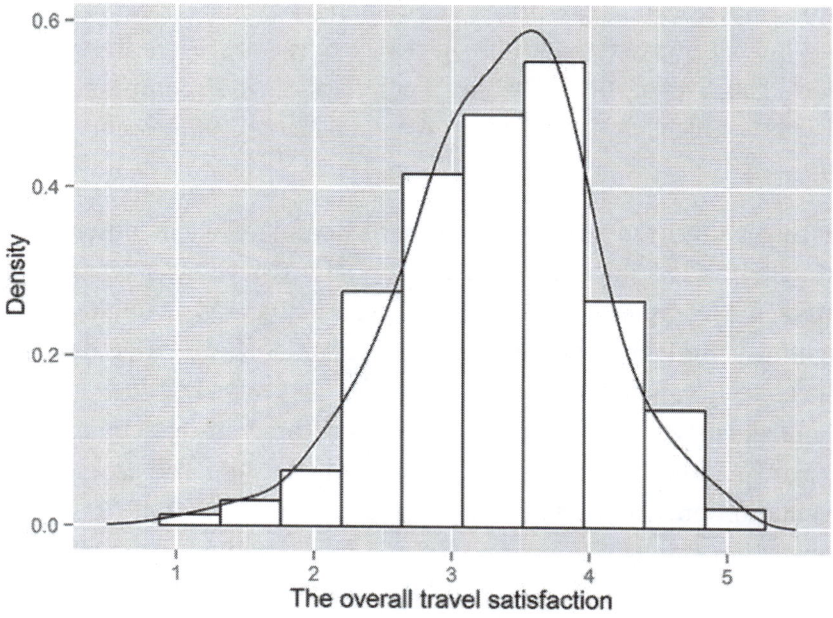

Figure 1. The histogram plot of the overall travel satisfaction superimposed with a density curve.

Table 1. Survey questions on overall travel satisfaction

Survey questions			Measurement
The overall degree of travel satisfaction	How well do you satisfy with ...	1. Public transport convenience 2. Traffic congestion conditions 3. Commuting convenience 4. Recreation-related travel convenience 5. Convenience to the city center	1 = very unsatisfied 2 = unsatisfied 3 = normal 4 = satisfied 5 = very satisfied

Much research has attempted to examine how subjectively experienced satisfaction correlates with various factors, such as sociodemographic attributes, density and diversity variables, accessibility to services, and social context. Conclusions remain equivocal (e.g., Ettema et al. 2012; Schwanen and Wang 2014). In our travel satisfaction model, the covariates are broadly divided into three categories. The first category includes a series of life circumstance variables or individual sociodemographic attributes, such as age, gender, and monthly income (Table 2). These variables are widely believed to have influence on overall life satisfaction, well-being, or happiness (e.g., Easterlin 2001; Ballas and Tranmer 2012). Second, a couple of travel-related variables, such as commuting time and transportation modes, are included in the travel satisfaction model. As the commute to and from work accounts for a significant part of daily travel, it might play an important role in overall travel satisfaction (Ettema et al. 2012). Moreover, we also take into account the interaction effects between commuting time and transportation modes on travel satisfaction.

Finally, a set of locational variables is included in the model to measure multiple dimensions of urban form, including population density, accessibility to public transit, proximity to green space, availability of various recreational facilities, and accessibility to the city center (Table 2). Drawing on these diversified locational variables can help us extend the understanding of how geographical factors affect individuals' subjective travel satisfaction. Furthermore, we investigate whether and how the effects of locational variables vary across districts in urban Beijing, with particular interest in the proximity effects of green parks and subway stations, which are allowed to be different in each geographical area. This is partly because spatial heterogeneity in the capitalization of access to green parks and subway stations into land values in the study area has been recently acknowledged (Harris, Dong, and Zhang 2013; Wu and Dong 2014). Moreover, although subway service expansion is currently the focus of municipal policies for encouraging public transport in Beijing, a further exploration of spatially varying effects of proximity to subway stations could be beneficial to public transit development (e.g., identifying particular districts with transport infrastructure priority) and transport policy evaluation (Ma, Mitchell, and Heppenstall 2015).

Table 2. Summarizing the travel satisfaction data used in the analysis

Variable names	Definition	Mean/proportions	*SD*
Dependent variable			
Travel	The overall travel satisfaction	3.333	0.698
Individual-level covariates			
Sociodemographic variables			
Income_1	Below 3,000 RMB	25.9%	
Income_2	3,000–4,999 RMB	38.7%	
Income_3	5,000–9,999 RMB	27.1%	
Income_4	Above 10,000 RMB	8.3%	
Age_below30	Age below 30	43.4%	
Age_30–39	Age between 30 and 39	23.8%	
Age_40–49	Age between 40 and 49	22.3%	
Age_above50	Age above 50	10.6%	
Female	Male as base category	50.2%	
Commute-related characteristics			
Commuting time	Log of resident's commuting time (minutes)	3.423	0.719
Commuting mode			
Public	Public transport such as bus and subway	46.8%	
Car	Private car	12.5%	
Others	Walk, cycling, and other modes	40.8%	
Locational variables			
Dist_subway	Log of distance to the nearest subway stations	7.087	0.977
Dist_recreation	Log of distance to the nearest recreational facilities (e.g., sport centers)	6.628	0.938
Dist_park	Log of distance to the nearest green parks	7.667	0.851
Dist_CBD	Log of distance to city center (central business district)	9.162	0.679
District-level variables			
Population density	Population density of each district (1,000 persons/km^2)	29.204	43.306

Note: RMB = Chinese Yuan Renminbi.

3.2 Travel Satisfaction Model Specification

The empirical travel satisfaction equation to estimate is as follows:$\text{Travel}_{i,j} = \alpha X'_{i,j} + \gamma L'_{i,j} + \delta P'_{i,j} + \beta_j A'_{i,j} + \epsilon_{i,j}$

$$\boldsymbol{\beta}_j = \left[\boldsymbol{\beta}_{1,j}, \boldsymbol{\beta}_{2,j}\right]' = \left[\boldsymbol{\beta}_1, \boldsymbol{\beta}_2\right]' + \left[u_{1,j}, u_{2,j}\right]';$$
$$\boldsymbol{u} \sim MVN(0, \Omega); \varepsilon \sim MVN\left(0, \sigma_\varepsilon^2 I\right), \tag{25}$$

where Travel represents travel satisfaction; *i* and *j* are individual and district indexes as in Equation 1; *X* represents sociodemographic variables including age, gender, and income (Table 2); *L* refers to locational variables of proximity to center city and recreational facilities whose effects are assumed to be fixed across districts; and *P* represents the urban form indicator (i.e., population density) at the district level. Parameters $\boldsymbol{a}$, $\boldsymbol{\gamma}$, and δ are fixed regression coefficients that we seek to estimate. *A* refers to proximity to green parks and subway stations and $\boldsymbol{\beta}_j$ are the associated regression coefficients that are assumed to be varied across districts. The fixed part (mean) of $\boldsymbol{\beta}_j$ is $[\beta_1, \beta_2]$ and the random part is specified by $[u_1, u_2]$, which follows a multivariate normal distribution with mean $\boldsymbol{0}$ and precision matrix Ω. Different formations of the precision matrix Ω from nonspatial random slope MLM, MLM-MLCAR, and MLM-MICAR were discussed earlier.

3.3 Model Comparisons

Comparing the performance of nonspatial random slope MLM, MLM-MLCAR, and MLM-MICAR, we adopt two commonly used Bayesian procedures: the deviance information criterion (DIC; Spiegelhalter et al. 2002) and the pseudo-Bayes factor (PsBF) calculated by an approximation to marginal likelihoods of two competing models (Neelon, O'Malley, and Normand 2010; Congdon 2014). The DIC is calculated as the sum of the posterior mean of the deviance (twice the negative log-likelihood of a model) and the number of effective model parameters. As a rule of thumb, if two competing models differ in DIC by more than three, the model with smaller DIC is regarded as a better fitting (Spiegelhalter et al. 2002). The Bayes factor (BF), obtained as the ratio of marginal likelihoods for two models, is another popular way to compare two competing models, which provides the evidence in the data that favors one model relative to another (Kass and Raftery 1995).

Following Neelon, O'Malley, and Normand (2010), to overcome the great computational difficulties of BF (calculation of marginal likelihoods of a model), we use the PsBF—an approximation of the BF through a cross-validation estimate of the marginal likelihood of a model (Gelfand and Dey 1994) to compare the performance of different models. The cross-validation predictive density for observation *i* is expressed as (Neelon, O'Malley, and Normand 2010)

$$f\left(y_{i,j}|\boldsymbol{y}_{[-(i,j)]}\right) = \int f\left(y_{i,j}|\boldsymbol{\theta},\boldsymbol{y}_{[-(i,j)]}\right)\pi(\boldsymbol{\theta}|\boldsymbol{y}_{[-(i,j)]})d\boldsymbol{\theta}, \qquad (26)$$

where $\boldsymbol{y}_{[-(i,j)]}$ denotes the vector of outcome with the $y_{i,j}$ deleted and θ represents the model parameters to estimate. The quantity $f(y_{i,j} \mid \boldsymbol{y}_{[-(i,j)]})$ is called the conditional predictive ordinate (CPO) for the *i*th observation in district *j*, of which the Monte Carlo estimate is given by

$$\widehat{\mathrm{CPO}}_{i,j} = \mathrm{R}/\sum_{r} 1/f(y_{i,j}|\boldsymbol{\theta}^{(r)}), \qquad (27)$$

where $\boldsymbol{\theta}^{(r)}$ refers to the parameter vector sampled at the lth iteration l = 1, ..., R (following a burn-in period) and $f(y_{i,j} \mid \boldsymbol{\theta}^{(r)})$ is the likelihood of observation $[i, j]$ evaluated at iteration l. The product of CPOs for each observation is called the pseudo-marginal likelihood (PsML; Gelfand and Dey 1994), and the ratio of PsMLs of two competing models gives a PsBF (Congdon 2014).

All competing models are coded using the R language. Diffuse or quite noninformative priors were placed on model parameters: fixed regression coefficients $\boldsymbol{\beta} \sim MVN\ (\mathbf{0}, 1000^*\boldsymbol{I}_{18})$, the precision matrix $\Gamma \sim dwish\ (\boldsymbol{I}_2, 2)$, and the individual level variance $\sigma_e^2 \sim IG\ (0.01, 0.01)$. Initial values for each model parameter were drawn from their corresponding probability distribution. For each of these models, statistical inferences were based on two MCMC chains, each of which consisted of 60,000 iterations with a burn-in period of 10,000. We further retained every tenth sample to reduce autocorrelation in each MCMC chain. MCMC diagnostics including trace plots and Brooks–Gelman–Rubin scale reduction statistics (Brooks and Gelman 1998) indicated rapid convergence of our samplers and efficient mixing of chains for each model under study. Figure 2 provides post-burn-in trace plots for several representative model parameters from MLM-MLCAR: fixed regression coefficients including the proximity to nearest green park (Dist_park) and also the proximity to nearest subway station (Dist_subway), the individual-level variance parameter Sigma2e, the spatial correlation parameter λ, and variance parameters of two sets of random effects (Var_Dist_park and Var_Dist_subway). The two much overlapped trajectory lines show convergence and efficient mixing of chains (Figure 2). The mean of the lag-5 autocorrelations for parameters in MLM-MLCAR was 0.003, ranging from −0.026 to 0.063, and the 97.5 percent quantiles of the Brooks–Gelman–Rubin statistics were all less than 1.01, indicating good convergence of each chain (Neelon, O'Malley, and Normand 2010).

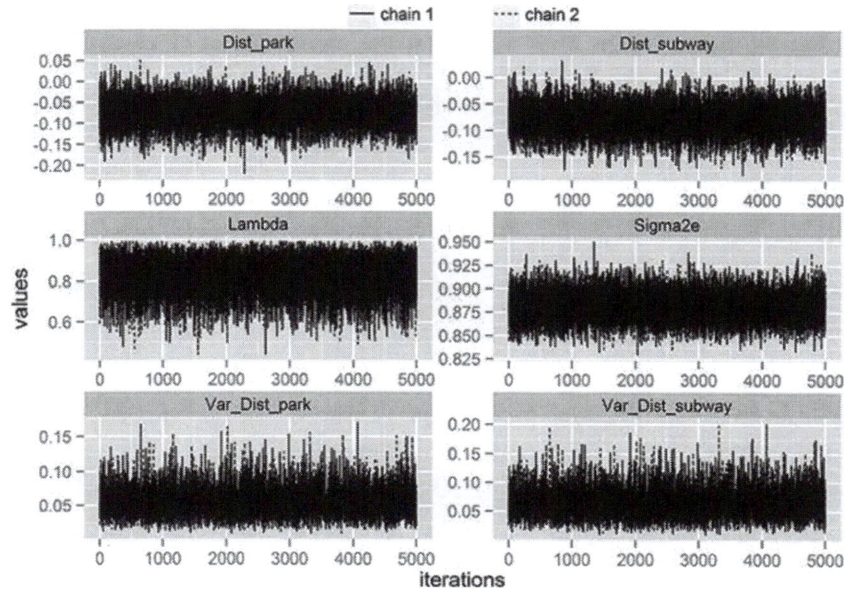

Figure 2. Trace plots based on two Markov chain Monte Carlo chains for six representative parameters from multilevel modeling-multivariate Leroux conditional autoregressive (MLM-MLCAR) model.

Model comparison results for the three models are provided in Table 3. All of the quantities are calculated based on the 10,000 post-burn-in samples from two MCMC chains. As shown in Table 3, the MLM-MLCAR produces the best model fit according to both the DIC and PsML statistics, compared to MLM-MICAR and nonspatial random slope MLM. Using the PsBF statistic, we find that the data strongly favor the MLM-MLCAR against nonspatial random slope MLM and MLM-MICAR with odds of 15.8 and 20.6, respectively (Kass and Raftery 1995). A better model fit of MLM-MLCAR against MLM-MICAR indicates the benefit of allowing the spatial correlation parameter to be estimated through data rather than defined a priori (equal to one in MLM-MICAR). The comparison between nonspatial random slope MLM and MLM-MICAR also supports this argument.

Table 3. Model fit comparisons using metrics of DIC, PsML, and PsBF

	DIC	PsML	PsBF (in favor of MLM-MLCAR)
MLM_MLCAR	17,602.0	–8,806.2	
MLM_MICAR	17,608.7	–8,809.2	
Nonspatial MLM	17,608.7	–8,809.0	
MLM_MLCAR/ MLM_MICAR	–6.7	3.0	20.6 (strong)
MLM_MICAR/ nonspatial MLM	–6.7	2.8	15.8 (strong)

Note: DIC = deviance information criterion; PsML = pseudo-marginal likelihood; PsBF = pseudo-Bayes factor; MLM = multilevel modeling; MLCAR = multivariate Leroux conditional autoregressive; MICAR = multivariate intrinsic conditional autoregressive.

4. RESULTS

Table 4 presents estimation results from MLM-MLCAR and nonspatial random slope MLM for a comparison. As shown later, posterior median and 95 percent credible intervals of each model parameter are produced. The estimate of the spatial correlation parameter in MLM-MLCAR is about 0.84 with a 95 percent credible interval [0.62, 0.97], indicating a fairly strong spatial dependence in each set of district-level random effects. We also investigate whether the district-level random effects of Dist_subway and Dist_park are spatially independent (as assumed to be so in nonspatial random slope MLM) by using the Moran's *I* statistic based on the same spatial weights matrix *W* used in MLM-MLCAR. The resultant Moran coefficient for random effects of Dist_subway is 0.144 with a *p* value equal to <0.001, and the Moran coefficient for random effects of Dist_park is 0.127 with a *p* value equal to <0.01. This demonstrates the existence of positive spatial dependence in the estimated random effects from nonspatial random slope MLM, which is in contradiction with the core model assumption of independence of random effects. Although there are no qualitative differences in fixed effect estimation between two models, MLM-MLCAR seems to be more efficient than nonspatial random slope MLM, as most of the 95 percent credible intervals of covariates are narrower than those from nonspatial random slope MLM.

Table 4. Summarizing the estimation results from the MLM-MICAR and MLM

	MLM-MLCAR			Nonspatial MLM		
	Posterior median	2.5%	97.5%	Posterior median	2.5%	97.5%
			Fixed effect estimates			
Intercept	1.546	0.848	2.250	1.479	0.712	2.268
Income_1	−0.117	−0.176	−0.057	−0.115	−0.177	−0.055
Income_3	0.066	0.009	0.124	0.068	0.009	0.126
Income_4	0.187	0.096	0.274	0.186	0.094	0.276
Age_below30	0.055	−0.003	0.115	0.056	−0.003	0.114
Age_40-49	−0.095	−0.161	−0.027	−0.095	−0.162	−0.026
Age_above50	−0.176	−0.271	−0.088	−0.172	−0.263	−0.079
Female	0.004	−0.043	0.052	0.005	−0.040	0.051
Commuting time	−0.046[a]	−0.096	0.003	−0.047	−0.098	0.004
Car	0.492	0.044	0.917	0.489	0.041	0.929
Public	0.791	0.508	1.085	0.787	0.494	1.094
Commuting time × Car	−0.158	−0.285	−0.027	−0.158	−0.287	−0.029
Commuting time × Public	−0.218	−0.298	−0.138	−0.217	−0.301	−0.134
Dist_subway	−0.078	−0.133	−0.023	−0.079	−0.132	−0.024
Dist_recreation	−0.009	−0.051	0.033	−0.010	−0.055	0.034
Dist_park	−0.076	−0.141	−0.008	−0.069	−0.137	−0.001
Dist_CBD	−0.110	−0.250	0.029	−0.093	−0.216	0.024
Population density	0.015	−0.065	0.094	0.017	−0.081	0.105
Lambda	0.840	0.620	0.970	N/A	N/A	N/A
Random effect estimates						
Sigma2e	0.881	0.852	0.913	0.879	0.850	0.909
*Var*_Dist_subway	0.059	0.025	0.117	0.028	0.015	0.052
*Var*_Dist_park	0.050	0.022	0.099	0.024	0.013	0.044
*Cov*_ Dist_park/Dist_park	−0.050	−0.103	−0.020	−0.023	−0.045	−0.012
*Corr*_ Dist_park/Dist_park	−0.926	−0.965	-0.822	−0.901	−0.947	−0.814

Note: MLM = multilevel modeling; MLCAR = multivariate Leroux conditional autoregressive; CBD = central business district.

[a]Although the main effect of commuting time is not statistically significant at the 95 percent credible level, it is significant at the 90 percent credible level with an interval of [−0.087, −0.005].

4.1 Travel Satisfaction and Life Circumstance

The estimates from MLM-MLCAR demonstrate that most of these life circumstance variables are significantly correlated with individuals' travel satisfaction (Table 4). For example, monthly income is significantly and positively associated with travel satisfaction. People with low-level income tend to have lower travel satisfaction, whereas high-level income tends to increase subjective travel satisfaction of residents. This broadly supports previous studies that have reported a significant impact of household income on life satisfaction or well-being, although the correlation is not linear (Easterlin 2001; Clark et al. 2008).

Distinctness is also found between different age cohorts. Older people are significantly associated with lower level of travel satisfaction, whereas young people (aged thirty and below) tend to score highest in satisfaction with travel. This is possibly because people aged forty and older are more likely to be involved in housework (e.g., child care and family errands) and recreational activities and thus they are more likely to be exposed to poor travel conditions in urban Beijing (Ma et al. 2014). Regarding the gender effect, there is no significant difference between men and women's travel satisfaction, which is in line with prior studies reporting similar satisfaction for gender (e.g., Ettema et al. 2012).

4.2 Travel Satisfaction and Commuting Behavior

Commute to and from work accounts for an important part of daily life, and it has a significant and negative effect on travel satisfaction. The MLM-MLCAR results show that with commuting time increasing, people's satisfaction with travel decreases while everything else equal, and this association is significant at the 90 percent credible level (Table 4). Commuting mode is also very significantly correlated with overall travel satisfaction. Compared to other travel modes such as walking or cycling, people commuting by car or public transport (i.e., bus and subway) tend to have higher travel satisfaction. The relationship between commuting behavior and travel satisfaction is complicated, however, by the inclusion of interaction effects between commuting time and modes, which are all statistically significant in the model.

As shown in Table 4, the negative regression coefficients of the interaction terms (Commuting time × Car and Commuting time × Public) suggest that with commuting time increasing, people's subjective satisfaction with travel by car and public transit tends to decrease. When commuting times for car or public transport surpass twenty-three or fifty-eight minutes, respectively, people's travel satisfaction with car or public transport is lower than that with other commuting modes, such as walking or cycling. From our survey data, we find that the average commuting time for the base category (e.g., walking and cycling) is approximately twenty minutes, which indicates that people are more likely to have higher satisfaction when they travel by car or public transport, rather than other commuting modes.

4.3 Travel Satisfaction and Geographical Context

The modeling results show that, of the locational variables, proximity to green parks and accessibility to subway stations are significantly associated with travel satisfaction. For instance, with close proximity to green parks at their residence, people tend to visit parks more often and they have higher travel satisfaction than their counterparts. Similarly, higher subway accessibility significantly increases residents' travel satisfaction, possibly due to the fact that the subway is fast, inexpensive, and uncongested in Beijing. People living in neighborhoods with higher population density or proximity to various recreational facilities also have higher travel satisfaction, although their correlations are not significant at the 95 percent credible level. Although proximity to city center is usually considered a proxy of employment accessibility, its correlation with travel satisfaction is insignificant. This is possibly because Beijing has undergone rapid urbanization, industrial decentralization, and residential suburbanization since the 1990s. High-tech industry zones and housing were established mainly in the suburbs, whereas employment opportunities for tertiary industries remained in the inner city (Ma et al. 2014). The emerging subcenters and coexistence of diversified land use configuration cause a complicated urban spatial structure in Beijing and also make it necessary to examine travel satisfaction under various geographical contexts.

Regarding the random effect estimates of locational variables, Figure 3 illustrates the spatially varying effects of proximity to subway stations in urban Beijing. The break points in Figure 3 correspond to the lower quartile, median, and upper quartile of estimated posterior means of the district-level effect of proximity to subway stations. Clusters with large effects of being close to subway stations (large negatives) are identified, including areas to the southeast of the central business district (CBD), areas near Zhongguancun (a subcenter in Beijing), and areas in the northwest suburb of Beijing. It suggests that increasing the subway accessibility, particularly for the districts around employment centers (e.g., the CBD and Zhongguancun) and large "bedroom" communities (e.g., Hui Long Guan in the northwest suburbs), will increase travel satisfaction of residents in these areas substantially. This is possibly because higher subway accessibility significantly increases commuting satisfaction on a typical workday, as a job–housing spatial mismatch exists in Beijing (Ma et al. 2014). This analysis, for the first time, investigates spatially varying effects of subway accessibility on travel satisfaction, identifies the particular districts with subway development priorities, and suggests useful solutions to improving people's well-being in Beijing.

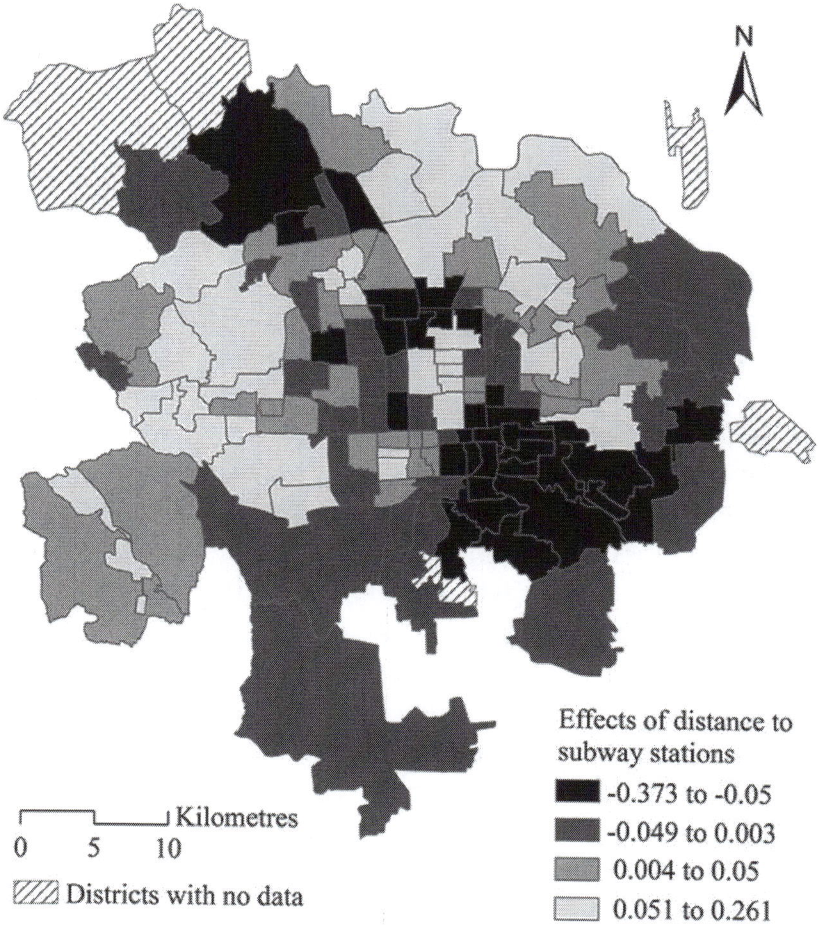

Figure 3. District-level effects of proximity to subway stations using estimates from multilevel modeling-multivariate Leroux conditional autoregressive (MLM-MLCAR) model.

Figure 4 illustrates the spatially varying effects of proximity to green parks, with break points at the lower quartile, median, and upper quartile of estimated posterior means at the district level. It also shows that the marginal effect of proximity to green parks on travel satisfaction varies across districts, with the high effect of proximity to green parks predominately distributed in the northern area of Beijing. Although the employment subcenters are mostly located in northern Beijing, better access to green space substantially increases travel satisfaction of residents in northern districts, possibly due to the higher satisfaction with recreational travels. In addition, the random effect estimates of proximity to green parks and subway stations from nonspatial random slope MLM and MLM-MLCAR are illustrated and compared in Figure 5, with a scatter plot superimposed by a 45-degree dashed line. By and large, it presents a good correspondence in estimates between these two models, with simple Pearson correlation coefficients for estimated effects of proximity to green parks and subway stations being 0.835 and 0.839, respectively. It also shows that the large positive effects of these two locational variables are reduced in MLM-MLCAR, whereas the small negative values are raised.

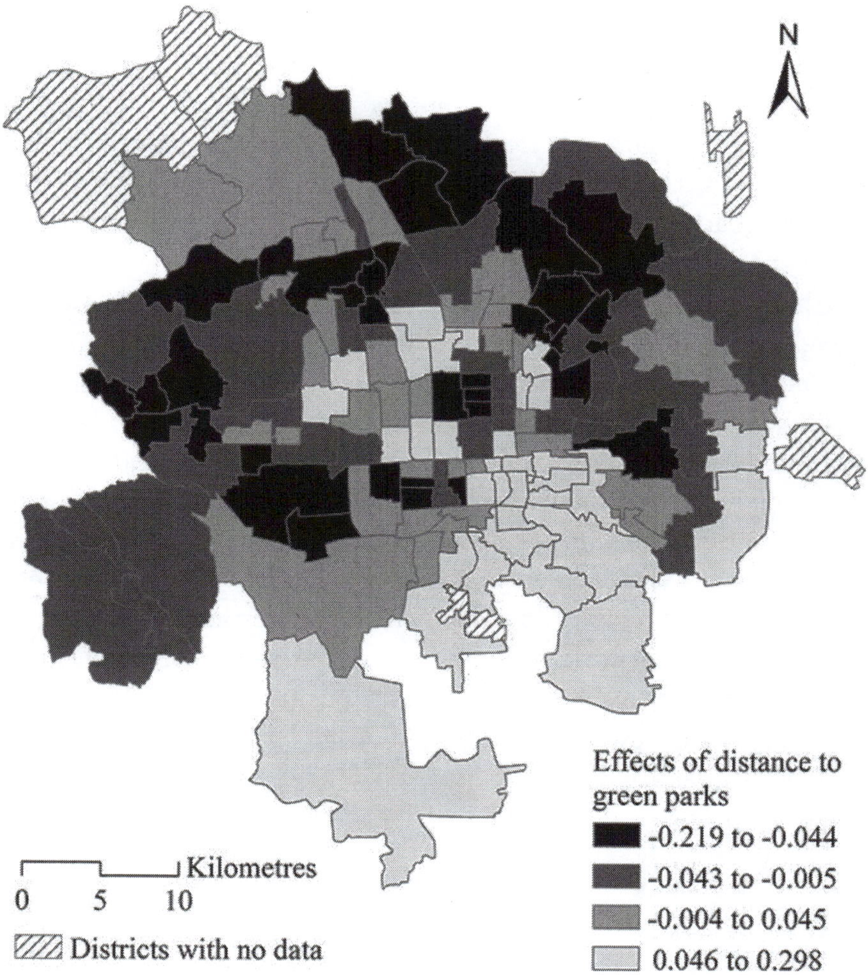

Figure 4. District-level effects of proximity to green parks using estimates from multilevel modeling-multivariate Leroux conditional autoregressive (MLM-MLCAR) model.

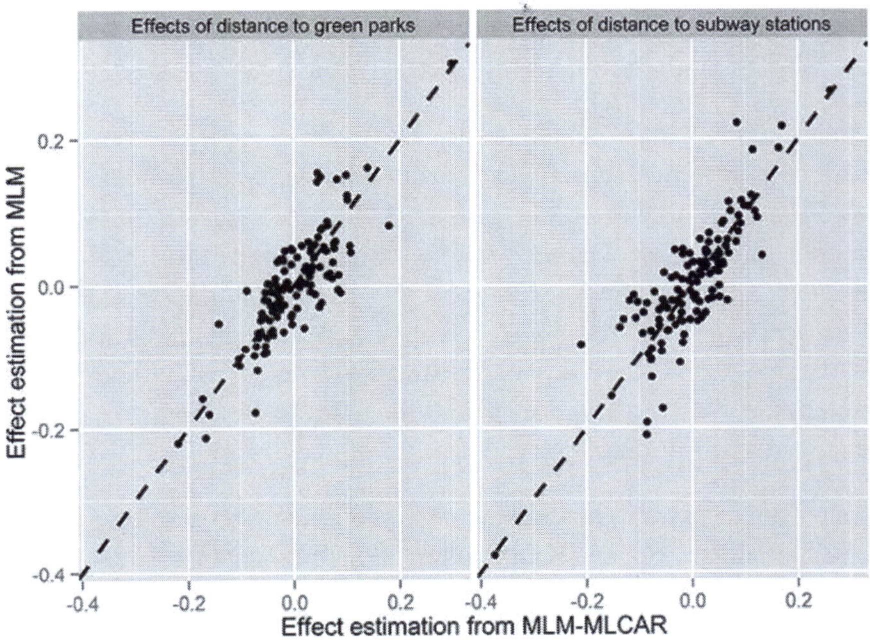

Figure 5. Comparing effect estimates f proximity to green parks and subway stations from multilevel modeling-multivariate Leroux conditional autoregressive (MLM-MLCAR) and nonspatial multilevel modeling (MLM).

5. CONCLUSION

This study presents a new modeling approach through the development of a spatial random slope MLM that incorporates the spatial dependence effects into standard nonspatial MLM. The motivation is that for many geographically clustered survey data, the crucial assumption of independence among the higher level spatial units underlying the standard nonspatial random slope MLM is challengeable. The geographical closeness of areal units is likely to induce spatial dependence in geographical-level random effects, which means that the contextual effects at higher level in random slope MLM might be dependent across space.

Building on an MCAR model, we define the area-level random effects as a complex and correlated spatial process, and propose a complex MLM-MLCAR approach. The proposed approach can simultaneously model the spatial dependence effect across space and the within-area correlations between different sets of random effects. Given the proliferation of geographically grouped data on individuals and their extensive use in social, health, and environmental research, we anticipate that the developed approach could be useful in a wide range of applications.

Using Beijing as a case study, we use the new method to investigate the spatial variations of people's subjective travel satisfaction and its determinants. Model results show that life circumstance variables including income and age and commute-related characteristics such as commuting time and travel mode choices are significantly associated with travel satisfaction. Regarding the locational variables, better access to subway stations and green parks is significantly correlated with higher levels of travel satisfaction. Furthermore, the effects of proximity to subway stations and green parks exhibit a fairly strong spatial pattern in Beijing, with high and low values clustering together, respectively. This could be due to the relatively large spatial correlation parameter identified in the model.

One potential issue not addressed here regards the sensitivity of the identified district-level effect, due to the notable issue of uncertain geographic context (Kwan 2012). The problem might exist in our case study, as districts might not be the true geographical context that influences individuals' subjective travel satisfaction. We hope, however, that using the developed hybrid spatial MLM that allows the outcomes to be not only influenced by their immediate geographical context but also by neighboring contexts, random effect estimates at the higher level can be more robust.

Although spatial dependence is considered globally in our model, we find that some districts that are geographical neighbors can have different random effect estimates (Figures 3 and 4). Our next step is to further incorporate a localized perspective of spatial dependence into the model using the approach developed by Lee and Mitchell (2013). A further extension of our method would allow for nonspatial dependence between aerial units. Pryce (2013), for example, argued that Euclidean distance or contiguity might not be the only way that dependence occurs between aerial units—perceived substitutability; social networks or communication and transport links might lead to some units being "close" even if they are spatially distant. In such circumstances, the conditional autoregressive component of the model could be constructed in substitutability or network space rather than Cartesian space (Łaszkiewicz, Dong, and Harris 2014).

ACKNOWLEDGMENTS

The authors are grateful for the comments of three reviewers and the editor, which have greatly improved the content of this article. They also would like to thank Professor Bill Browne and Dr. George Leckie from the Centre for Multilevel Modeling (CMM), University of Bristol for providing valuable comments on the previous draft of the article. Thanks are also due to Professor Wenzhong Zhang and Dr. Jianhui Yu from the Institute of Geographical Sciences and Natural Resources Research, Chinese Academy of Sciences for providing the data used in the study.

FUNDING

1. For each of the spatial multilevel models implemented, the convergence of the MCMC sampler is diagnosed using the CODA package (Plummer et al. 2006) in R. In terms of efficient computation, a crucial part is updating the spatially random effect, a *JP* by 1 (384 by 1 in this case) vector based on its full posterior conditional distribution. We take advantage of a desirable characteristic of the GMRFs, the sparsity of their precision matrix, and therefore some fast sparse matrix Cholesky factorization can be carried out. More specifically, a canonical parameterization of the posterior distribution of the spatial random effect is used to draw samples of them via a very useful and computationally efficient function in an R package, SPAM, created by Furrer and Sain (2010). Details on fast sampling algorithms for GMRFs are provided in Rue and Held (2005) and Furrer and Sain (2010). It takes about four minutes for the MCMC sampler of the MLM-MLCAR model to produce 10,000 samples on a laptop with an Intel Core 2.5 GHz processor. Before applying the code to the travel satisfaction data, we conducted a series of simulation studies with known data generating processes and using the geography (data structure) of the travel data to test the code. The results show that spatial multilevel models can accurately retrieve the true model parameters. The R code for implementing the spatial multilevel models and the simulation study results are available on request.

2. Based on estimates from MLM-MLCAR in Table 4, the marginal effects of commuting by car and public transports (with other transport modes as baseline category) are 0.492 + (−0.158) × Commuting time and 0.791 + (−0.218) × Commuting time, respectively. Equating the two marginal effects to zero (and exponentiating the solutions) gives the commuting time thresholds in the main text.

REFERENCES

1. Anselin, L. 1988. *Spatial econometrics: Methods and models*. Dordrecht, The Netherlands: Kluwer Academic.
2. Arcaya, M., M. Brewster, C. Zigler, and S. V. Subramanian. 2012. Area variations in health: A spatial multilevel modeling approach. *Health and Place* 18:824–31.
3. Assuncao, R. M. 2003. Space varying coefficient models for small area data. *Environmentrics* 14:453–73.
4. Ballas, D., and M. Tranmer. 2012. Happy people or happy place?: A multilevel modelling approach to the analysis of happiness and well-being. *International Regional Science Review* 35:70–102.
5. Baltagi, B. H., B. Fingleton, and A. Pirotte. 2014. Spatial lag models with nested random effects: An instrumental variable procedure with an application to English house prices. *Journal of Urban Economics* 80:76–86.
6. Banerjee, S., B. P. Carlin, and A. E. Gelfand. 2004. *Hierarchical modeling and analysis for spatial data*. Boca Raton, FL: Chapman and Hall/CRC.
7. Besag, J., J. C. York, and A. Mollie. 1991. Bayesian image restoration, with two principle applications in spatial statistics. *Annals of the Institute of Statistical Mathematics* 43:1–59.
8. Brooks, S. P., and A. Gelman. 1998. General methods for monitoring convergence of iterative simulations. *Journal of Computational and Graphical Statistics* 7:434–55.
9. Browne, W. J., and H. Goldstein. 2010. MCMC sampling for a multilevel model with nonindependent residuals within and between cluster units. *Journal of Educational and Behavioral Statistics* 35:453–73.
10. Browne, W. J., H. Goldstein, and J. Rasbash. 2001. Multiple-membership multiple-classification (MCMC) models. *Statistical Modelling* 1:103–24.
11. Chaix, B., J. Merlo, and P. Chauvin. 2005. Comparison of a spatial approach with the multilevel approach for investigating place

effects on health: The example of healthcare utilisation in France. *Journal of Epidemiology and Community Health* 59:517–26.

12. Chaix, B., J. Merlo, S. V. Subramanian, J. Lynch, and P. Chauvin. 2005. Comparison of a spatial perspective with the multilevel analytical approach in neighborhood studies: The case of mental and behavioural disorders due to psychoactive substance use in Malmö, Sweden, 2001. *American Journal of Epidemiology* 162:171–82.

13. Chib, S., and E. Greenberg. 1995. Understanding the Metropolis–Hastings algorithm. *The American Statistician* 49:327–35. [Taylor & Francis Online]

14. Clark, A. E., E. Diener, Y. Georgellis, and R. E. Lucas. 2008. Lags and leads in life satisfaction: A test of the baseline hypothesis. *The Economic Journal* 118:222–43.

15. Cliff, A., and J. K. Ord. 1981. *Spatial processes: Models and applications.* London: Pion.

16. Congdon, P. 2014. *Applied Bayesian modeling*. 2nd ed. Boca Raton, FL: Chapman & Hall/CRC.

17. Corrado, L., and B. Fingleton. 2012. Where is the economics in spatial econometrics. *Journal of Regional Sciences* 52:210–39.

18. Cressie, N. 1993. *Statistics for spatial data*. Rev. ed. New York: Wiley.

19. Dong, G. P., and R. Harris. 2015. Spatial autoregressive models for geographically hierarchical data structures. *Geographical Analysis* 47:173–91.

20. Dong, G. P., R. Harris, K. Jones, and J. Yu. 2015. Multilevel modelling with spatial interaction effects with application to an emerging land market in Beijing, China. *PLoS ONE* 10 (6): e0130761.

21. Easterlin, R. A. 2001. Income and happiness: Towards a unified theory. *The Economic Journal* 111:465–84.

22. Ettema, D., M. Friman, T. Garling, L. E. Olsson, and S. Fujii. 2012. How in-vehicle activities affect work commuters' satisfaction with public transport. *Journal of Transport Geography* 24:215–22.

23. Fotheringham, A. S., C. Brunsdon, and M. Charlton. 2002. *Geographically weighted regression: The analysis of spatially varying relationships.* Chichester, UK: Wiley.

24. Furrer, R., and S. Sain. 2010. Spam: A sparse matrix R package with emphasis on MCMC methods for Gaussian Markov random fields. *Journal of Statistical Software* 36 (10): 1–25.

25. Gelfand, A. E. 2012. Hierarchical modeling for spatial data problems. *Spatial Statistics* 1:30–39.

26. Gelfand, A. E., S. Banerjee, C. F. Sirmans, Y. Tu, and S. E. Ong. 2007. Multilevel modeling using spatial processes: Application to the Singapore housing market. *Computational Statistics and Data Analysis* 51:3567–79.

27. Gelfand, A. E., and D. Dey. 1994. Bayesian model choice: Asymptotics and exact calculations. *Journal of the Royal Statistical Society: Series B* 56:501–14.

28. Gelfand, A. E., and P. Vounatsou. 2003. Proper multivariate conditional autoregressive models for spatial data analysis. *Biostatistics* 4:11–25.

29. Gelman, A., B. Carlin, H. S. Stern, and D. B. Rubin. 2004, *Bayesian Data Analysis,* 2nd ed. Chapman & Hall/CRC.

30. Goldstein, H. 2003. *Multilevel statistical methods.* 3rd ed. London: Arnold.

31. Haining, R. 2003. *Spatial data analysis: Theory and practice.* Cambridge, UK: Cambridge University Press.

32. Harris, R., G. P. Dong, and W. Zhang. 2013. Using contextualized geographically weighted regression to model the spatial heterogeneity of land prices in Beijing, China. *Transactions in GIS* 17:901–19.

33. Kass, R. E., and A. E. Raftery. 1995. Bayes factors. *Journal of the American Statistical Association* 90:773–95.

34. Kwan, M. P. 2012. The uncertain geographic context problem. *Annals of the Association of American Geographers* 102:958–68.

35. Łaszkiewicz, E., G. P. Dong, and R. Harris. 2014. The effect of omitted spatial effects and social dependence in the modelling of household expenditure for fruits and vegetables. *Comparative Economic Research* 17:155–72.
36. Lee, D. 2011. A comparison of conditional autoregressive models used in Bayesian disease mapping. *Spatial and Spatio-temporal Epidemiology* 2:79–89.
37. Lee, D., and R. Mitchell. 2013. Locally adaptive spatial smoothing using conditional auto-regressive models. *Journal of the Royal Statistical Society: Series C* 62:593–608.
38. Leroux, B., X. Lei, and N. Breslow. 1999. Estimation of disease rates in small areas: A new mixed model for spatial dependence. In *Statistical models in epidemiology, the environment, and clinical trials*, ed. M. Halloran and D. Berry, 135–78. New York: Springer-Verlag.
39. LeSage, J. P., and R. K. Pace. 2009. *Introduction to spatial econometrics*. Boca Raton, FL: CRC Press/American Association of Geographers.
40. Ma, J., A. Heppenstall, K. Harland, and G. Mitchell. 2014. Synthesising carbon emission for mega-cities: A static spatial microsimulation of transport CO_2 from urban travel in Beijing. *Computers, Environment and Urban Systems* 45:78–88.
41. Ma, J., G. Mitchell, and A. Heppenstall. 2015. Exploring transport carbon futures using population microsimulation and travel diaries: Beijing to 2030. *Transportation Research Part D* 37:108–22.
42. MacNab, Y. C. 2011. On Gaussian Markov random fields and Bayesian disease mapping. *Statistical Methods in Medical Research* 20:49–68.
43. Martinez-Beneito, M. A. 2013. A general modelling framework for multivariate disease mapping. *Biometrika* 100:539–53.
44. Meng, B., L. Zheng, and H. Yu. 2011. Commuting time change and its influencing factors in Beijing. *Progress in Geography* 30:1218–24.

45. Morenoff, J. D. 2003. Neighborhood mechanisms and spatial dynamics of birth weight. *American Journal of Sociology* 108:976–1017.
46. Neelon, B., R. Anthopolos, and M. L. Miranda. 2014. A spatial bivariate probit model for correlated binary data with application to adverse birth outcomes. *Statistical Methods in Medical Research* 23:119–33.
47. Neelon, B., A. O'Malley, and S. Normand. 2010. A Bayesian model for repeated measures zero-inflated count data with application to outpatient psychiatric service use. *Statistical Modelling* 10:421–39.
48. Owen, G., R. Harris, and K. Jones. 2015. Under examination: Multilevel models, geography and health research. *Progress in Human Geography*. Advance online publication. doi:10.1177/0309132515580814
49. Plummer, M., N. Best, K. Cowles, and K. Vines. 2006. CODA: Convergence diagnosis and output analysis for MCMC. *R News* 6:7–11.
50. Pryce, G. 2013. Housing submarkets and the lattice of substitution. *Urban Studies* 50 (13): 2682–99.
51. Raudenbush, S. W., and A. S. Bryk. 2002. *Hierarchical linear models, applications, and data analysis methods*. New York: Sage.
52. Rue, H., and L. Held. 2005. *Gaussian Markov random fields: Theory and applications*. New York: Chapman and Hall/CRC.
53. Savitz, N. V., and S. W. Raudenbush. 2009. Exploiting spatial dependence to improve measurement of neighborhood social processes. *Sociological Methodology* 39:151–83.
54. Schwanen, T., and D. Wang. 2014. Well-being, context, and everyday activities in space and time. *Annals of the Association of American Geographers* 104 (4): 833–51.
55. Spiegelhalter, D., N. Best., B. P. Carlin., and A. Van Der Linde. 2002. Bayesian measures of model complexity and fit (with discussion). *Journal of the Royal Statistical Society: Series B* 64:583–639.

56. Wall, M. M. 2004. A close look at the spatial structure implied by the CAR and SAR models. *Journal of Statistical Planning and Inference* 121:311–24.

57. Wu, W., and G. Dong. 2014. Valuing the "green" amenities in a spatial context. *Journal of Regional Sciences* 54:569–85.

58. Zhang, W., W. Yin, and J. Zhang. 2006. *A study of liveable cites in Beijing, China*. Beijing, China: Beijing Social Sciences Academic Press.

CHAPTER 4

Dirichlet Compound Multinomials Statistical Models

Paola Cerchiello, Paolo Giudici

Department of Economics and Management, University of Pavia, Pavia, Italy

ABSTRACT

This contribution deals with a generative approach for the analysis of textual data. Instead of creating heuristic rules for the representation of documents and word counts, we employ a distribution able to model words along texts considering different topics. In this regard, following Minka proposal (2003), we implement a Dirichlet Compound Multinomial (DCM) distribution, then we propose an extension called sbDCM that takes explicitly into account the different latent topics that compound the document. We follow two alternative approaches: on one hand the topics can be unknown, thus to be estimated on the basis of the data, on the other hand topics are determined in advance on the basis of a predefined ontological schema. The two possible approaches are assessed on the basis of real data.

KEYWORDS

Textual Data Analysis; Mixture Models; Ontology Schema; Reputational Risk

1. INTRODUCTION

With the rapid growth of on-line information, text categorization has become one of the key techniques for handling and organizing data in textual format. Text categorization techniques are an essential part of text mining and are used to classify new documents and to find interesting information contained within several on-line web sites. Since building text classifiers by hand is difficult, time-consuming and often not efficient, it is worthy to learn classifiers from experimental data. In this proposal we employ a generative approach for the analysis of textual data. In the last two decades many interesting and powerful contributions have been proposed. In particular, when coping with the text classification task, a researcher has to face the well-known problem of polysems (multiple senses for a given words) and synonyms (same meaning for different words). One of the first effective model able to solve those issues is represented by Latent semantic analysis (LSA) [1]. The basic idea is to work at a semantical level by reducing the vector space through Singular Value Decomposition (SVD), producing not sparse occurrence tables that help in discovering associations between documents. In order to establish a solid theoretical statistical framework in this context, in [2] a probabilistic version of LSA (pLSA) has been proposed, also known as the aspect model, rooted in the family of latent class models and based on a mixture of conditionally independent multinomial distributions for the couple words-documents. The intention from the introduction of pLSA was to offer a formal statistical framework, helping the parameter interpretation issue as well. By the way the goal was achieved only partially, in fact the multinomial mixtures, which components can be interpreted as topics, offer a probabilistic justification at words but not at documents level. In fact the latter are represented merely as list of mixing proportions derived from mixture components. Moreover, the multinomial distribution presents as many values as there are in the training documents and therefore it learns topic mixture on those trained documents. The extension to previously unseen documents is not appropriate since there can be new topics. In order to overcome the asymmetry between words and documents and to produce a real generative model, [3] proposed the LDA (Latent Dirichlet Allocation). The idea of such new approach emerges from the concept of exchangeability for the words in a document that unfolds in the "bag of words" assumption: the order of words in a text is not important. In fact the LDA model is able to capture either the words or documents exchangeability unlike LSA and pLSA. On the other hand LDA is a generative model in any sense since it posits a Dirichlet distribution over documents in the corpus, while each topic is drawn from a Multinomial distribution over words. However note that [4] in 2003 have shown that LDA and pLSA are equivalent if the latter is under a uniform Dirichlet prior distribution. Obviously LDA does not solve all the issues. The main restriction embedded in LDA approach and due to the Dirichlet distribution, is the assumption of independence among topics. The immediate consequence was to tackle the issue by introducing the Correlated Topic Model (CTM), as proposed in [5]. CTM introduces correlations among topics by replacing the Dirichlet random variable with the logistic normal distribution. Unlike LDA, CTM presents a clear complication in terms of inference and parameter estimation since the logistic normal distribution and the Multinomial are not conjugate. To bypass the problem, the most recent alternative is represented by the Independent Factor Topic Models (IFTM) introduced in [6]. Such proposal makes use of latent variable model approach to detect hidden correlations among topics. The choice to explore the latent model world allows to choose among several alternatives ranging from the type of relation, linear or not linear, to the type of prior to be specified for the latent source. For sake of completeness is important to mention another interesting research path focusing on the burstiness phenomenon, that is the tendency of rare words, mostly, to appear in burst. The above mentioned generative models are not able to capture such peculiarity, that instead is very well modelled by the Dirichlet Compund Multinomial model (DCM). Such distribution was introduced by statisticians [7] and has been widely employed by other sectors like bioinformatics [8] and language engineering [9]. An important contribution in the context of text classification was brought by [10] and [11] that profitably used DCM as a bag-of-bags-of-words generative process. Similarly to LDA, we have a Dirichlet random variable that generates a Multinomial random variable for each document from which words are drawn. By the way, DCM cannot be considered a topic model in a way, since each document derives specifically by one topic. That is the main reason why [12] proposed a natural extension of the classical topic model LDA by plugging into it the DCM distribution and obtaining the so called DCMLDA. Following this line of thinking, we move from DCM approach and we propose an extension of the DCM, called "semanticbased Dirichlet Compound Multinomial" (sbDCM), that permits to take latent topics into account. The paper is organized as follows: in Section 2 we show the Dirichlet Compound Multinomial (DCM) model, in Section we propose an extension of the DCM, called "semanticbased Dirichlet Compound Multinomial" (sbDCM), in Section 4 we show how to estimate the parameters of the different models. Then, in Section 5 we assess the predictive performance of the two distributions by using seven different classifiers. Finally we show the different classification performance according to the knowledge on the topics T (known or unknown).

2. BACKGROUND: THE DIRICHLET COMPOUND MULTINOMIAL

The DCM distribution is a hierarchical model: on one hand, the Dirichlet random variable is devoted to model the Multinomial word parameters q; on the other hand, the Multinomial variable models the word count vectors $\overline{x}$ comprising the document. The distribution function of the DCM mixture model is:

$$p(\overline{x}|\alpha)=\int_{\theta}p(\overline{x}|\theta)p(\theta|\alpha)\mathrm{d}\theta \quad . (1)$$

where $p(\overline{x}|\theta)$ is the Multinomial distribution:

$$p(\overline{x}|\theta)=\frac{n!}{\prod_{w=1}^{W}x_w}\prod_{w=\backslash}^{W}\theta_w^{x_w}. \quad (2)$$

in which $\overline{x}$ is the words count vector, x_w is the count for each word and q_w the probability of emitting a word w_i; therefore a document is modelled as a single set of words ("bag-of-words"). The Dirichlet distribution $p(\theta|\alpha)$ is instead parameterized as:

$$p(\theta|\alpha)=\frac{\Gamma\left(\sum_{w=1}^{W}\alpha_w\right)}{\prod_{w=1}^{W}\Gamma(\alpha_w)}\prod_{w=1}^{W}\theta_w^{\alpha_w-1}. \tag{3}$$

with $\alpha=\{\alpha_w\}$ the Dirichlet parameter vector for words, as consequence the whole set of words ("bag-of-bags") is modelled. Thus a text (a document in a set) is modelled as "bag-of-bags-of-words". Developing the previous integral we obtain:

$$p(\theta|\alpha)=\frac{n!}{\prod_{w=1}^{W}(x_w)}\frac{\Gamma\left(\sum_{w=1}^{W}\alpha_w\right)}{\Gamma\left(\sum_{w=1}^{W}(x_w+\alpha_w)\right)}\prod_{w=1}^{W}\frac{\Gamma(x_w+\alpha_w)}{\Gamma(\alpha_w)} \tag{4}$$

In Figure 1 we report the graphical representation of the DCM model. From another point of view, each Multinomial is linked to specific sub-topics and makes, for a specific document, the emission of some words more likely than others. Instead the Dirichlet represents a general topic that compounds the set of documents and thus the DCM could be also described as "bag-of-scaleddocuments". The added value of the DCM approach consists in the ability to handle the "burstiness" of a rare word without introducing heuristics [13]. In fact, if a rare word appears once along a text, it is much more likely to appear again.

When we consider the entire set of documents (D), where each document is independent and identified by its count vector, $\left(D=\left\{\overline{x}_1,\overline{x},\cdots,\overline{x}_N\right\}\right)$, the likelihood of the whole documents set (D) is

$$p(D|\alpha)=\prod_{d=1}^{N}p(\overline{x}_d|\alpha)$$

$$=\prod_{d=1}^{N}\left(\frac{\Gamma\left(\sum_{w=1}^{W}\alpha_w\right)}{\Gamma\left(x_d+\sum_{w=1}^{W}\alpha_w\right)}\prod_{w=1}^{W}\frac{\Gamma(x_w+\alpha_w)}{\Gamma(\alpha_w)}\right) \tag{5}$$

where x_d is the sum of the counts of each word in the document d-th (åx_{dw}) and x_{dw} the count of word w-th for the document d-th. Thus the log-likelihood is:

$$\log\left(p(D|\alpha)\right)$$

$$=\sum_{d=1}^{N}\log\Gamma\left(\sum_{w=1}^{W}\alpha_w\right)-\log\Gamma\left(x_w+\sum_{w=1}^{W}\alpha_w\right)$$

$$+\sum_{d=1}^{N}\sum_{w=1}^{W}\log\Gamma(x_{dw}+\alpha_w)-\log\Gamma(\alpha_w) \tag{6}$$

The parameters can be estimated by a fixed-point iteration scheme, as described in Section 4.

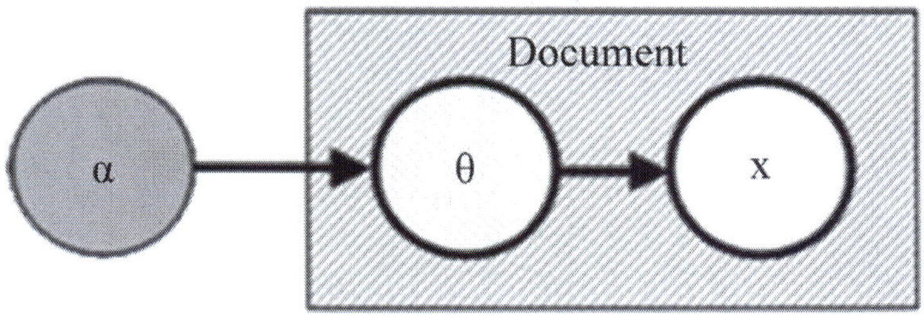

Figure 1. Hierarchical representation of DCM model.

3. A SEMANTIC-BASED DCM

As explained in Section 2, we have a coefficient a_w for each word compounding the vocabulary of the set of documents which is called "corpus". The DCM model can be seen as a "bag-of-scaled-documents" where the Dirichlet takes into account a general topic and the Multinomial some specific sub-topics. Our aim in this contribution is to build a framework that allows us to insert specifically the topics (known or unknown) that compound the document, without losing the "burstiness" phenomenon and the classification performance. Thus we introduce a method to link the a coefficients to the hypothetic topics, indicated with $\beta = \{\beta_i\}$, by means of a function $\alpha = F(\beta)$ which must be positive in b since the Dirichlet coefficients are positive. Note that usually $\dim(b) \ll \dim(a)$ and, therefore, our proposed approach is parsimonious. Substituting the new function into the integral in Equation (1), the new model is:

$$p(\bar{x}|\beta) = \int_\theta p(\bar{x}|\theta)p(\theta|F(\beta))\mathrm{d}\theta, \quad (7)$$

We have considered as function F(b) a linear combination based on a matrix D and the vector $\overline{\beta}$. D contains information about the way of splitting among topics the observed count vectors of the words contained in a diagonal matrix A and $\overline{\beta}$ is a vector of coefficient (weights) for the topics. More specifically we assume that:

$$A = \begin{pmatrix} w_1 & & \\ & \ddots & \\ & & w_W \end{pmatrix}, D = \begin{pmatrix} d_{11} & \cdots & d_{1T} \\ \vdots & \ddots & \vdots \\ d_{W1} & \cdots & d_{WT} \end{pmatrix},$$

$$\beta = \begin{pmatrix} \beta_1 \\ \vdots \\ \beta_t \\ \vdots \\ \beta_T \end{pmatrix}, D^* = A \times D \tag{8}$$

$$F\left(\overline{\beta}\right) = D^* \times \overline{\beta} = \alpha = \begin{pmatrix} \alpha_1 \\ \vdots \\ \alpha_t \\ \vdots \\ \alpha_T \end{pmatrix} \tag{9}$$

Note that:

- $\alpha_w = \sum_t^T d_t^* \beta_t$ with T the number of Topics;

- $d^*_{wt} = w_w \times d_{wt}$;
- d_{wt} is the coefficient for word w-th used to define the degree of belonging to topic t-th and by which a portion of the count of word w-th is assigned to that particular topic t-th.

By substituting this linear combination into Equation (4), we obtain the same distribution but with the above mentioned linear combination for each a:

$$p(\bar{x}|\beta) = \frac{n!}{\prod_{w=1}^{W}(x_w)} \frac{\Gamma\left(\sum_{w=1}^{W}\sum_{t}^{T} d^*_{wt}\beta_t\right)}{\Gamma\left(\sum_{w=1}^{W}\left(x_w + \sum_{t}^{T} d^*_{wt}\beta_t\right)\right)} \prod_{w=1}^{W} \frac{\Gamma\left(x_w + \sum_{t}^{T} d^*_{wt}\beta_t\right)}{\Gamma\left(\sum_{t}^{T} d^*_{wt}\beta_t\right)} \quad (10)$$

This model is a modified version of the DCM, henceforth semantic-based DCM (sbDCM), and the new loglikelihood for the set of documents becomes:

$$\log(p(D|\beta)) = \sum_{d=1}^{N} \log\Gamma\left(\sum_{w=1}^{W}\sum_{t}^{T} d^*_{wt}\beta_t\right) - \log\Gamma\left(x_w + \sum_{w=1}^{W}\sum_{t}^{T} d^*_{wt}\beta_t\right) + \sum_{d=1}^{N}\sum_{w=1}^{W}\left(\log\Gamma\left(x_{dw} + \sum_{t}^{T} d^*_{wt}\beta_t\right) - \log\Gamma\left(\sum_{t}^{T} d^*_{wt}\beta_t\right)\right) \quad (11)$$

In Figure 2 we report the graphical representation of the new model where the a's are substituted by a function of the b's. An important aspect of the proposed approach is represented by the number T of topics to be inserted into the sbDCM that can be:

- Unknown, thus to be estimated on the basis of the available data.
- A priori known (i.e. fixed by field experts).

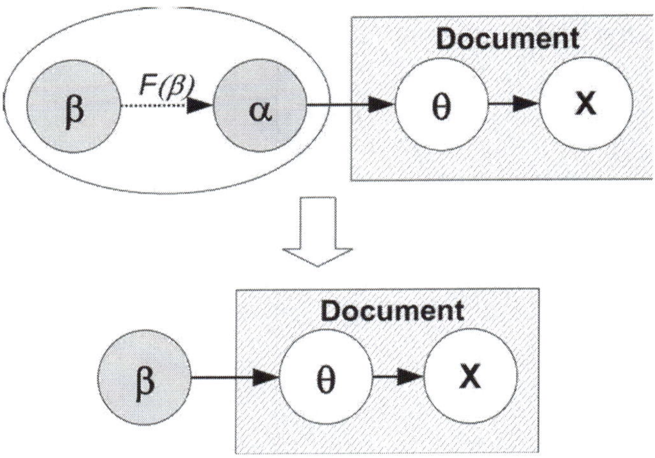

Figure 2. Hierarchical representation of sbDCM model.

The first case will be treated in Section 5.1, in particular since it is not always possible to know in advance the number of latent topics present in a corpora, it becomes very useful to build a statistical methodology for discovering efficiently and consistently a suitable number of topics. In this case the number of topics T can be considered a random variable. Thus, we use a segmentation procedure to group the words in order to create groups of words sharing common characteristics that can be considered as latent topics. The analysis is completed by choosing the best number of groups and a distance matrix is used to set the membership percentage (d_{wt}) of each word to each latent topic. The second case will be treated in Section 5.2 and proposes to exploit the sbDCM model by employing a priori knowledge based on ontological schemas that describe the relations among concepts with regards to the general topics of the corpora. The ontology structure provides the set of relations among the concepts to which can be associated a certain number of key words, by a field expert(s). Thus, we want to use the classes of a given ontology and the associated key words to define in advance the number T of topics.

4. PARAMETERS ESTIMATION

There are several methods to maximize the log-like- lihood and to find the parameters associated with a DCM: simplified Newton iteration, the Expected Maximization method and the maximization of the simplified likelihood (called "leave-one-out" likelihood, LOO). Among them, the most general and flexible algorithm is the Expected Maximization (EM). The EM algorithm is an iterative procedure able to compute the maximum-likelihood estimates whenever data are incomplete or they are considered complete but not observable (latent variables) [14].

In general, considering $p(X,Z|\theta)$ the joint probability distribution (or probability mass function) where X is the incomplete (but observable) data set and Z the missing (or latent) data, we can calculate the complete maximum likelihood estimation of model parameters q by means of the EM algorithm. We obtain an overall complexity reduction for what concerns the calculation of observed maximum-likelihood (ML) estimates. The goal is achieved by alternating the expectation (E-step) of the likelihood with the latent variables as if they are observed and the maximization (M-step) of the likelihood function found in the E-step. With the result of the M-step we update the new parameters q_{new} that are used again in the cycle, by starting from the E-step, until we reach a fixed degree of approximation e of the observed likelihood which increases step by step moving towards the maximum (that could be local) [15]. The EM can be built in different ways. One possibility is to see the EM as a lower bound maximization where we alternate the E-step to calculate an approximation of the lower bound for the log-likelihood and maximize it in the M-step until a stationary point (zero gradient) is reached. If we are able to find a lower bound for the log-likelihood we can maximize it via a fixed-point iteration; in fact it is the same principle of considering the EM as a lower bound maximization. In our context, for the DCM, the lower bound with $\log\left(p\left(D|\alpha\right)\right)$ is the following quantity:

$$\alpha_w^{k+1} = \alpha_w^k \frac{\sum_w^k \left[\psi\left(x_{dw} + \alpha_w^k\right) - \psi\left(\alpha_w^k\right)\right]}{\sum_d^N \left[\psi\left(x_d + \sum_w^W \alpha_w^k\right) - \psi\left(\sum_w^W \alpha_w^k\right)\right]} \quad (12)$$

this allows us to use a fixed point iteration whose steps are:

$$\alpha_w^{k+1} = \alpha_w^k \frac{\sum_w^k \left[\psi\left(x_{dw} + \alpha_w^k\right) - \psi\left(\alpha_w^k\right)\right]}{\sum_d^N \left[\psi\left(x_d + \sum_w^W \alpha_w^k\right) - \psi\left(\sum_w^W \alpha_w^k\right)\right]} \quad (13)$$

where x_d is the sum of the counts of each word in the document d-th $\left(\sum x_{dw}\right)$, x_{dw} the count of word w-th for the document d-th and α_w^k the Dirichlet coefficient for word w at the k-th step. The algorithm is stopped when a degree of approximation e is reached. The iteration starts with a_w equals to the occurrence percentage of the word w-th in the corpus. The estimated parameters, as said before, have an important characteristic: they follow the "burstiness" phenomenon of words. In fact the smaller a_w is, the more "burstiness" effect is contained within a word, as revealed in Section 5.

In the case of sbDCM by considering the new loglikelihood, Equation (11), we can use the same lower bound employed before. The only modification lays on the substitution of the a coefficients with the linear function of b, as in Equation (9), and thereby the new fixed point iteration step is:

$$\beta_t^{k+1} = \beta_t^k \frac{\sum_d^N \left[\sum_w^W \left(d_{wt}^* \psi \left(x_{dw} + \sum_t^T d_{wt}^* \beta_t^k \right) \right) - \sum_w^W \left(d_{wt}^* \right) \psi \left(\sum_t^T d_{wt}^* \beta_t^k \right) \right]}{\sum_d^N \left[\sum_w^W \left(d_{wt}^* \right) \psi \left(x_{dw} + \sum_w^W \sum_t^T d_{wt}^* \beta_t^k \right) - \sum_w^W \left(d_{wt}^* \right) \psi \left(\sum_w^W \sum_t^T d_{wt}^* \beta_t^k \right) \right]} \qquad (14)$$

As before we stop the iteration at a fixed degree of approximation and the coefficients d_{wt} are those described in Section 3. The new a's maintain the words burstiness, as we shall show in Section 5 and they are used to classify the document by employing a Naive Bayes classifier. For our applications, in the next section we have used for both models a value of e of 10^{-10}. The iteration starts with b_t equals to the percentage of each single cluster obtained by the grouping analysis of vocabulary words.

5. MODEL PERFORMANCE

In this section we describe the evaluation of the different classifiers by using the parameters estimated from the DCM distribution. Thus, our training data set is compound of 6436 documents with a vocabulary (already filtered and stemmed) of 15,655 words so we have to estimate 15,655 a parameters. The a parameter is able to model the "burstiness" of a word. In fact, the smaller the a parameters are, the more bursty the emission of words is. This phenomenon is characteristic of rare words, therefore a coefficients are, on average, smaller for less counted words. The average value of the overall a parameters is 0.0342, the standard deviation is 0.1087 and the maximum and minimum values are respectively 6.6074 and 0.0025. Once the coefficient vector of a's is obtained we employ seven different classifiers, three of which are described in [13] (normal (N), complement (C) and mixed (M). The remaining ones are proposed as the appropriate combination of the previous ones, in order to improve their characteristics. Those new classifiers are set in function of the number of words that a test-document has in common with the set of documents that compound a class; in this way we create a classifier in function of the number of words in common. Thus we analyze the following additional classifiers: COMPLEMENT + MIXED + NORMAL (CMN), COMPLEMENT + NORMAL (CN), COMPLEMENT + MIXED (CM), MIXED + NORMAL (MN).

In order to evaluate the classification performance we employ three performance indexes:

- Ind1: The proportion of true positive over the total number of test-documents:

$$\left(\sum_{d=1}^{D}\frac{TP_d}{D}\right)\times 100;$$

- Ind2: The proportion of classes that we are able to classify:

$$\left(\sum_{c=1}^{C}\frac{I_c}{C}\right)\times 100;$$

where I_c is an indicator that we set 1 if at least one document of the class is classified correctly, otherwise we set 0.

- Ind3: The proportion of true positive within each class over the number of test documents present in the class:

$$\left(\frac{1}{C}\sum_{c=1}^{C}\frac{TP_c}{M_c}\right)\times 100;$$

where M_c is the number of test-documents in each class, TP_c is the number of true positive in the class and C the number of topics (46).

For the four combined classifiers such indexes have been calculated by varying the number of words in common between the test document and the class. In particular for our test we have used three different thresholds for the number of words (n): 15, 10 and 5. For example, we indicate with the initials CM_n the classification rule that employs classifier C when the number of common words are less or equal to n and classifier M when the number of words in common is more than n. Instead, the initial $CMN_{n.m}$ identifies the using of classifier C until n, the classifier N over m and the classifier M between n and m. For the data at hand the number of words in common between the two sets (training and evaluation set) varies between 1 and 268. The above mentioned combination is based on the following idea: if the number of words in common between the bag of words and the correct class is low, then the most information content is in the complement set. Otherwise the needed information is contained either in the normal set or in the complement one. Taking into account such consideration we have set up different combination and we concluded that the useful trade-off among classifiers is equal to 10 (Table 1).

As we can see the best classifiers are the mixed and the CM_{10} ones. They are able to classify respectively 1237 and 1238 over 1609 documents that are distributed not uniformly among classes (46). These classifiers are able to classify at least a document per class even if there are classes containing only 2 documents. Between them the CM_{10} classifier has index three slightly better than mixed one. The worse classifier, in this case, is the complement version alone.

Table 1. Classification results by varying cluster numbers and using matrix C.

Classifier	Measure	*sbDCM*5	*sbDCM*11	*sbDCM*17	*sbDCM*23	*sbDCM*46	*DCM*
	LL_t	−282,226	−265,250	−252,197	−247,125	−242,991	−222,385
	LL_e	−205,412	−205,614	−205,601	−205,597	−205,602	−205,286
	AIC_t	56,446	530,522	504,228	494,296	486,074	454,264
	AIC_e	410,834	411,250	411,236	411,240	411,296	420,066
Norm.	*Ind*1	68.13	68.13	67.95	68.19	68.25	68.78
\\	*Ind*2	97.83	97.83	97.83	97.83	97.83	97.83
\\	*Ind*3	62.32	62.32	62.15	62.32	62.32	61.61
Comp	*Ind*1	68.19	68.31	68.25	68.37	68.25	68.78
\\	*Ind*2	100	100.0	100.0	100.0	100.0	100.0
\\	*Ind*3	66.20	66.30	66.05	66.58	66.01	67.89
Mixed	*Ind*1	68.07	68.36	68.43	68.37	68.31	68.07
\\	*Ind*2	97.83	97.83	97.83	97.83	97.83	97.83
\\	*Ind*3	63.87	64.01	64.05	64.05	64.01	63.87

We now verify the goodness of the sbDCM models described in Section 3, to understand if we can insert latent topics into DCM by maintaining the burstiness and the same classification performance. Two kinds of matrixes have been used in the cluster procedure. One matrix contains the correlations among words C in the vocabulary and another G is constructed by calculating the Kruskal-Wallis index on the count matrix among words. The latter index g is defined as follows:

$$g = \frac{\frac{12}{N(N+1)}\sum_{i=1}^{K} n_i \left(\overline{r_i} - \frac{N+1}{2}\right)^2}{1 - \frac{\sum_{i=1}^{p}\left(c_i^3 - c_i\right)}{N^3 - N}} \qquad (15)$$

where n_i is the number of sample data, N the total observation number of the k samples, k the number of samples to be compared and $\overline{r_i}$ the mean rank of i-th group. The denominator of the index g is a correction factor needed when tied data are present in the data set, where p is the number of recurring ranks and c_i is the times the i-th rank is repeated. The index g depends on the differences among the averages of the groups $\overline{r_i}$ and the general average. If the samples come from the same population or from populations with the same central tendency, the arithmetic averages of the ranks of each group

$\overline{r_i} = \sum_j r_{ij} / n_i$ should be similar to each other and to the general average $(N+1)/2$ as well. The training dataset contains 2051 documents with a vocabulary of 4096 words for both approaches. The evaluation dataset (again the same for both models) contains

1686 documents which are distributed over 46 classes. In Tables 1 and 2 we report the results from the two tests obtained respectively by matrixes C and G and by varying the number of groups in the cluster. In the tables we indicate with LL_{in} the log-likelihood before the parameters updating and with LL_{out} after the iteration procedure which is stopped when the error e = 10^{-10} is reached. The same with AIC_{cin} and AIC_{cout} that is the corrected Akaike Information Criterion (AIC_c) before and after the uploading. The indexes Ind1, Ind2 and Ind3 have been described in Section 5.1.

As we can see in the two Tables 1 and 2, the percentages of correct classification (Ind1) are very close to the original ones with a parameter for each word (4096 parameters). Of course they depend on the type of classifier employed during the classification step. Considering both sbDCM and DCM, the differences produced by varying the number of groups are small. Moreover the AIC_c is always better in the new approach then considering each word as a parameter (DCM model). In particular for what concerns the approach based on the correlation matrix C (in Table 2) with 17 groups and on the Mixed classifier, it can predict correctly the 68.43% of documents. The log-likelihood and the AIC_c indexes along groups are quite similar, however the best value is obtained with 5 groups (respectively −205,412 and 410,834). Considering again the approach based on the correlation matrix C, we can conclude that, in terms of complexity expressed by the AIC index, the sbDCM approach whatever applied classifier is always better than the DCM. When we use matrix G (Table 2) the best classification performance is for the complement classifier based on 23 groups, with a percentage of 68.72%, a log-likelihood of −204,604, the AIC_c of 409,254. The best log-likelihood and AIC_c are for cluster with 46 groups (respectively −204,362 and 408,816). Even if the sbDCM distribution based on matrix G is not able to improve the classification performance of DCM, we can say that the sbDCM Index1 is always very close to the best one. In addition the new model is always better in terms of either AIC and log-likelihood indexes. Moreover, if we perform an asymptotic chisquared test (χ^2_{test}) considering the two cases (matrixes G and C) to decide whether the difference among loglikelihoods (LL), with respect to DCM, are significant (i.e. the difference is statistically meaningful if the $|LL_1 - LL_2|$ is greater than 6), we can see from Tables 1 and 2 the test with matrix G has the best performance.

Table 2. Classification results by varying cluster numbers and using matrix G.

Classifier	Measure	*sbDCM*5	*sbDCM*11	*sbDCM*17	*sbDCM*23	*sbDCM*46	*DCM*
	LL_i	−291,257	−283,294	−270,360	−266,453	−258,061	−222,385
	LL_o	−205,912	−204,647	−204,600	−204,604	−204,362	−205,286
	AIC_i	582,524	566,610	540,754	532,952	516,214	454,264
	AIC_o	411,834	409,316	409,234	409,254	408,816	420,066
Norm.	*Ind*1	67.83	67.71	67.47	67.42	67.65	67.66
\\	*Ind*2	97.83	97.83	97.83	97.83	97.83	97.83
\\	*Ind*3	62.02	61.73	61.45	61.43	61.55	61.61
Comp	*Ind*1	67.95	68.66	68.55	68.72	68.60	68.78
\\	*Ind*2	100	100.0	100.0	100.0	100.0	100.0
\\	*Ind*3	67.95	68.09	68.05	68.29	68.05	67.89
Mixed	*Ind*1	68.07	68.13	67.83	67.71	67.89	68.07
\\	*Ind*2	97.83	97.83	97.83	97.83	97.83	97.83
\\	*Ind*3	63.87	63.97	63.05	62.86	62.95	63.87

Performance of the Semantic-Based Dirichlet Compound Multinomial with T Known in Advance

A different approach needs to be assessed when the number of available topic T is known in advance. In fact a text corpora could be enriched by several descriptions of treated topics according to the knowledge of field experts. In more details, the analysis could be provided with a priori knowledge based on ontological schemas that describe the relations among concepts with regards to the general topics of the corpora. An ontology (from which an ontological schema is derived) is a formal representation of a set of concepts within a domain and the relationships between those concepts [16]. It provides a shared vocabulary, which can be used to model a domain, that is, the type of objects and/or concepts that exist, and their properties and relations. In Figure 3 we report an example of graphical representation of an ontological schema. For example, if a text set deals with reputational risk management for corporate institutions, an ontology can be created on the basis of the four categories of problems (internal processes, people, systems and external events) defined by Basel II Accords.

Hence we can suppose that some specific sub-topics and key words, such as the possible causes of reputetional losses, will be almost surely treated along the texts.

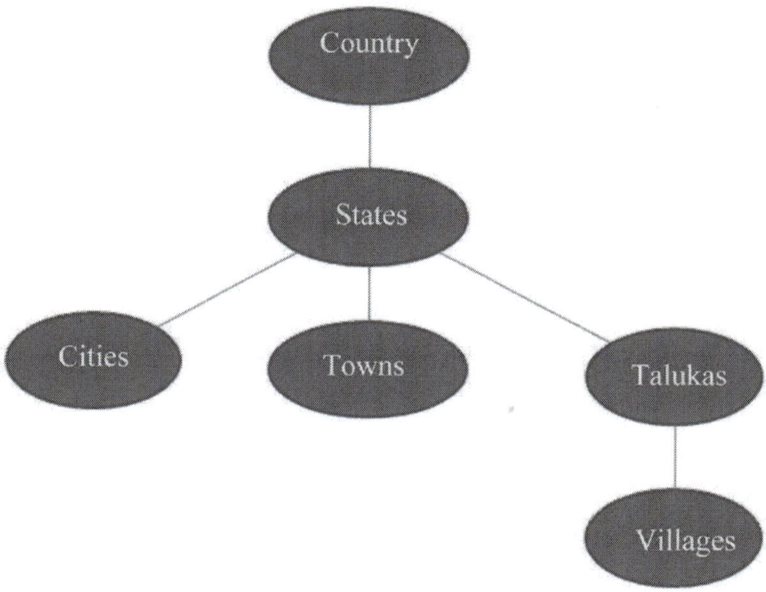

Figure 3. Example of ontological schema.

Thereby, the ontology structure provides the set of relations among the concepts to which can be associated, by a field expert(s), a certain number of key words. Thus, we want to use the classes of a given ontology and the associated key words to define in advance the number T of topics. The ontological schema which we refer to, for the data at hand, deals with the so called reputational risk [17]. It is not simple to define and consequently to measure and to monitor the reputation concept since it involves intangible assets such as: honor, public opinion, perception, reliability, merit. By the way, it is a matter of fact, that a bad reputation can seriously affect and condition the performance of a company. Moreover companies tend to act once the adverse event has occurred. According to such approach we can say that there is not a risk management activity, but only a crisis management. With regards to reputational risk, media coverage plays a key role in determining a company's reputation. This often occurs when a company reputation has been significantly damaged by unfair attacks from special interest groups or inaccurate reporting by the media. A detailed and structured analysis of what the media are saying is especially important because the media shape the perceptions and expectations of all the involved actors. Natural language processing technologies enable these services to scan a wide range of outlets, including newspapers, magazines, TV, radio, and blogs. In order to enable the application of the classification textual model sbDCM we have collaborated with the Italian market leader company in financial and economic communication, Sole24ORE. Sole- 24ORE team provided us with a set of 870 articles about Alitalia, an Italian flight company, covering a period of one year (Sept 07-Sept 08).

The 80% of the articles are used to train the model and the remaining 20% to validate the process. The objective is to classify the articles on the basis of the reputation ontology in order to understand the argument treated in the articles. The ontology classes used for the classification are:

- Identity: the perception that stakeholders have of the organization, person, product. It describes how the organization is perceived by the stakeholders.
- Corporate Image: the "persona" of the organitation. Usually for companies visibly manifested by way of branding and the use of trademarks and involves the mission and the vision. It involves brand value.
- Integrity: personal inner sense of "wholeness" deriving from honesty and consistent uprightness of character.
- Quality: the achievement or excellence of an entity. Quality is sometimes certificated by a third part.
- Reliability: ability of a system to perform/maintain its functions in routine and also in different hostile or/ and unexpected circumstances. It involves customer satisfaction and customer fidelitation.
- Social Responsibility: social responsibility is a doctrine that claims that an organization or individual has a responsibility to society. It involves foundation campaign and sustainability
- Technical Innovation: the Introduction of new technical products or services. Measure of the "RD orientation" of an organization (only for companies).
- Value For Money: the extent to which the utility of a product or service justifies its price.

Those classes define the concept of reputation of a company. To link the ontology classes to the textual analysis we use a set of key words for each class of the reputation schema. Since the articles are in Italian, the key words are in Italian as well. For example the concept of "Reliability" involves customer satisfaction and customer fidelitation and is characterized by the following set of key words: affidabilità, fiducia, consumatori, risorsa, organizzazione, commerciale, dinamicità, valore, mercato. On the basis of these key words, we perform a grouping analysis considering the 9 classes. From the clustering, we derive the matrix D. Empirical results show good performance in terms of correct classification rate. We obtain that, given 171 articles, we correctly classify the 68% of them.

6. CONCLUSIONS

This contribution has shown how to enrich the DCM model with a semantic extension. We also have proposed a method to insert latent topics within the "Dirichlet Compound Multinomial" (DCM) without losing the words "burstiness": we call such a distribution "semanticbased Dirichlet Compound Multinomial" (sbDCM). The approaches assessed depend on the knowledge about the topics T. In fact there can be two alternative contexts: on one hand the topics are unknown in advanced, thus to be estimated on the basis of data at hand. On the other hand a text corpora could be enriched by several descriptions of treated topics according to the experience of the field expert(s). Specifically, the analysis can be empowered with a priori knowledge based on ontological schemas that describe the relations among concepts with regards to the general class argument of the corpora. In order to insert topics we create a new coefficient vector b_t for each topic and later on we obtain the a parameters as a linear combination of them. The methodology is based on a matrix D containing the degree of membership of each word to a cluster (i.e. a topic) by using the cluster distance matrix. Then we split the words count vectors among latent topics and, by employing a fixed-point iteration, we generate the β coefficients representing the topics weights. In order to compare the two models DCM and sbDCM we have employed a "Naive Bayes Classifier" based on the estimated distributions as shown in [13]. Several classifiers have been proposed and tested, and among them the best performance is obtained by means of the "mixed formula" and "CM_{10}". Moreover, we run several tests to verify if the classification performance reached with an a for each word (DCM) is maintained or improved by the sbDCM.

Such an objective has been accomplished employing two different approaches. We propose two different methods to generate b parameters, one based on the correlation among words C and the second based on the Kruskal-Wallis index calculated on the words count matrix G. The results report that the test performances in terms of misclassification rate are quite close to each other and to the performance reached by the DCM. However the sbDCM distribution is able to obtain better results in terms of AIC and log-likelihood especially in the case of matrix G. Concluding, by using matrix D to describe how words count vectors can be split among topics, the b as weights for the topic and a as a linear combination we are able to obtain an optimal classification performance and to follow the burstiness.

ACKNOWLEDGEMENTS

This work has been supported by MUSING 2006 contract number 027097, 2006-2010 and FIRB, 2006-2009). The paper is the result of the close collaboration between the authors, however the paper has been written by Paola Cerchiello under the supervision of Paolo Giudici.

REFERENCES

1. S. Deerwester, S. Dumais, G. W. Furnas, T. K. Landauer and R. Harshman, "Indexing by Latent Semantic Analysis," Journal of the American Society for Information Science, Vol. 41, No. 6, 1990, pp. 391-407.

2. T. Hofmann, "Probabilistic Latent Semantic Indexing," Proceedings of Special Interest Group on Information Retrieval, New York, 1999, pp. 50-57.

3. D. M. Blei, A. Y. Ng and M. I. Jordan, "Latent Dirichlet Allocation," Journal of Machine Learning Research, Vol. 3, 2003, pp. 993-1022.

4. M. Girolami and A. Kaban, "On an Equivalence between PLSI and LDA," Proceedings of Special Interest Group on Information Retrieval, New York, 2003, pp. 433-434.

5. D. M. Blei and J. D. Lafferty, "Correlated Topic Models," Advances in Neural Information Processing Systems, Vol. 18, 2006, pp. 1-47.

6. D. Putthividhya, H. T. Attias and S. S. Nagarajan, "Independent Factor Topic Models," Proceeding of International Conference on Machine Learning, New York, 2009, pp. 833-840.

7. J. E. Mosimann, "On the Compound Multinomial Distribution, the Multivariate B-Distribution, and Correlations among Proportions," Biometrika, Vol. 49, No. 1-2, 1962, pp. 65-82.

8. K. Sjolander, K. Karplus, M. Brown, R. Hughey, A. Krogh, I. S. Mian and D. Haussler, "Dirichlet Mixtures: A Method for Improving Detection of Weak but Significant Protein Sequence Homology," Computer Applications in the Biosciences, Vol. 12, No. 4, 1996, pp. 327- 345.

9. D. J. C. Mackay and L. Peto, "A Hierarchical Dirichlet Language Model," Natural Language Engineering, Vol. 1, No. 3, 1994, pp. 1-19.

10. T. Minka, "Estimating a Dirichlet distribution," Unpublished Paper, 2003. http://research.microsoft.com/en-us/um/people/minka/papers/dirichlet/

11. R. E. Madsen, D. Kauchak and C. Elkan, "Modeling Word Burstiness Using the Dirichlet Distribution," Proceeding of the 22nd International Conference on Machine Learning, New York, 2005, pp. 545-552.

12. G. Doyle and C. Elkan, "Accounting for Burstiness in Topic Models," Proceeding of International Conference on Machine Learning, New York, 2009, pp. 281-288.

13. J. D. M. Rennie, L. Shih, J. Teevan and D. R. Karge, "Tackling the Poor Assumptions of Naive Bayes Text Classifier," Proceeding of the 20th International Conference on Machine Learning, Washington DC, 2003, 6 p.

14. A. P. Dempster, M. N. Laird and D. B. Rubin, "Maximum Likelihood from Incomplete Data via the EM Algorithm," Journal of the Royal Statistical Society, Series B, Vol. 39, No. 1, 1977, pp. 1-38.

15. D. Böhning, "The EM Algorithm with Gradient Function Update for Discrete Mixture with Know (Fixed) Number of Components," Statistics and Computing, Vol. 13, No. 3, 2003, pp. 257-265.

16. S. Staab and R. Studer, "Handbook on Ontologies, International Handbooks on Information Systems," 2nd Edition, Springer, Berlin, 2009.

17. P. Cerchiello, "Statistical Models to Measure Corporate Reputation," Journal of Applied Quantitative Methods, Vol. 6, No. 4, 2011, pp. 58-71.

CHAPTER 5

Modeling individual differences in randomized experiments using growth models: Recommendations for design, statistical analysis and reporting of results of internet interventions

Hugo Hesser

Department of Behavioural Sciences and Learning,
Linköping University, Sweden

ABSTRACT

Growth models (also known as linear mixed effects models, multilevel models, and random coefficients models) have the capability of studying change at the group as well as the individual level. In addition, these methods have documented advantages over traditional data analytic approaches in the analysis of repeated-measures data. These advantages include, but are not limited to, the ability to incorporate time-varying predictors, handle dependence among repeated observations in a very flexible manner, and to provide accurate estimates with missing data under fairly unrestrictive missing data assumptions. The flexibility of the growth curve modeling approach to the analysis of change makes it the preferred choice in the evaluation of direct, indirect and moderated intervention effects. Although offering many benefits, growth models present challenges in terms of design, analysis and reporting of results. This paper provides a nontechnical overview of growth models in the analysis of change in randomized experiments and advocates for their use in the field of internet interventions. Practical recommendations for design, analysis and reporting of results from growth models are provided.

KEYWORDS

Growth models; Structural equation modeling; Multilevel modeling; Internet interventions; Mediation; Moderation

1. INTRODUCTION

Despite considerable advances in applied statistics and methodology over the years, the randomized experiment is still the only established way to test cause-effect associations (Pearl, 2009). By randomizing participants to conditions and holding everything, save one variable (i.e., experimental manipulation), constant across conditions, the average difference between participants treated and controls is an unbiased estimate of the average casual effect of the treated (Rubin, 1974). A well-designed randomized experiment also allows for the exploration of how this direct effect between treatment and outcome arises (i.e., mediation/indirect effects), and under what circumstances or for whom (i.e., moderation) this effect is most pronounced (Baron and Kenny, 1986).

A number of factors can bias the estimate of the direct, indirect, and moderated effects in a randomized experiment. Modern statistical analyses of repeated-measures data offer potential remedies to common problems encountered in field experiments, for example, addressing problems with attrition/missing data, dependence among repeated observations, and statistical power. Methodological advances in the analysis of change also provide researchers with excellent analytic tools to answer important research questions regarding individual differences. These advances are most relevant for the statistical analysis of mediation and moderation in randomized experiments. Despite the benefits of these techniques, however, these methods also present specific and real challenges for researchers in terms of designing, analyzing and reporting results of treatment trials.

The field of internet interventions has grown rapidly over the past decade (Andersson, 2009 and Andrews et al., 2010). Numerous randomized controlled trials have been conducted to evaluate whether internet-delivered interventions produce beneficial outcomes relative to control conditions, other treatments, or delivery-formats for a wide variety of health issues (see for reviews, Andersson et al., 2011, Andersson et al., 2014, Andrews et al., 2010, Hedman, 2014 and Spek et al., 2007). There is clear heterogeneity, however, among studies on internet interventions in terms of the scientific methodology used (Barak et al., 2009, Danaher and Seeley, 2009 and Ritterband and Tate, 2009). Recently, in an attempt to raise the quality of research in the area, general research guidelines were developed that provide researchers with recommendations for how to best report studies within the field (Proudfoot et al., 2011); yet, to my knowledge, no guidelines exist that provide recommendations for how to appropriately design and analyze randomized experiments of internet interventions. Such guidelines can be of critical importance for the scientific progression of the field at large, as an appropriate methodology of experiments is at the heart of scientific claims and inferences (Mayo, 1996). Specifically, no guidelines have been developed that provide recommendations for how to make use of modern statistical techniques in the analysis of change in experiments of internet interventions. Perhaps as a consequence, coupled with the rapid growth of research in the area, the data analytic procedures used in clinical trials on internet interventions have varied between studies and some studies have used out-of date methods to analyze clinical trial data. This could lead to incorrect conclusions and in the long run be problematic for the field. In addition, studies within the field have yet to make full use of the advantages of modern data analytic techniques offer in the study of individual differences in change and correlates of individual change in experimental designs.

The overall purpose of this paper is to increase awareness of the advantages of modern data analytic procedures for repeated-measures data and to encourage their use over traditional methods in the field of internet interventions. The paper provides a nontechnical overview of growth curve models in the analysis of direct, indirect and moderated treatment effects and summarizes some recommendations from leading methodological authorities within the field of longitudinal data analysis. The objective is not to provide a detailed technical account of how to conduct these analyses. For this, readers are referred to the many accessible didactic articles and books available (Bollen and Curran, 2006, Hedeker and Gibbons, 2006, Kwok et al., 2008, Peugh and Enders, 2005, Singer, 1998, Singer and Willett, 2003 and Snijders and Bosker, 2012).

The structure of the paper is as follows. The article begins with a brief review of growth models in the analysis of change in randomized experiments. Here, I also focus on design issues of relevance when using growth models. An example is provided to illustrate how to implement and interpret a linear growth model in the context of a randomized controlled trial. I then provide some recommendations for modeling building issues concerning functional form of change, coding of time, and variance components. Next, I focus on a common problem in randomized experiments that can seriously bias inferences: missing data. In the context of Rubin's (1976) missing data theory, I discuss maximum likelihood estimation in growth models and argue for its use in clinical trials with incomplete data over other common ad hoc missing data handling techniques. In the subsequent section, I discuss how to take full advantage of the modeling capabilities of growth models in the statistical analysis of mediation and moderation in randomized experiments. The paper concludes with some recommendations for the reporting of results from growth models in studies on internet interventions.

2. MEASURE AND MODEL CHANGE IN RANDOMIZED EXPERIMENTS

When researchers conduct a randomized experiment to evaluate an outcome of an intervention they are, implicitly or explicitly, concerned with how that outcome changes over time. They might be interested in how a specific set of behaviors or symptomatology expressed by the individual at certain time point has changed in response to the intervention. Consequently, when evaluating the outcome of an intervention, they need to consider how to model and measure change. Historically, change has often been examined using one or two assessment points, such as in a pre- and post-treatment design (Francis et al., 1991). For such research designs, repeated measures analysis of variance (RM-ANOVA) or multiple regression analysis (including analysis of covariance which is a form of multiple regression) has frequently been employed. The primary purpose here is to study mean group differences. Individual deviations from these averages are treated as error variance, rather than reflecting the true amount of change that take place at the individual level (Duncan and Duncan, 2004). Although this design and data analytic choice can be defended, concerns from a conceptual and statistical standpoint have been raised by methodologists, in particular, related to the study of correlates of change (Rogosa and Willett, 1985). On a conceptual level, these models do not explicitly model individual change and change is viewed as incremental rather than as a continuous developmental process that unfolds over time (Francis et al., 1991, Rogosa, 1988 and Rogosa and Willett, 1985). On a statistical level, these models are concerning because of the (un-) reliability of the difference score, inability to adequately handle time-varying predictors, missing data as well as unbalanced data (Duncan and Duncan, 2004, Francis et al., 1991, Gueorguieva and Krystal, 2004, Kwok et al., 2008, Rogosa and Willett, 1985 and Willett and Sayer, 1994). Moreover, our data analytic approach must be able to handle nested-data structures and dependence among observations in repeated-measures data (i.e., observations are nested within individuals over time). Traditional data analytic approaches (e.g., RM-ANOVA) rest on strong data analytic assumptions regarding covariance structures for handling dependence among repeated-measurements (e.g., sphericity assumption) and if these assumptions are violated it will lead to incorrect decisions (Francis et al., 1991, Gueorguieva and Krystal, 2004 and Kwok et al., 2008).

An alternative approach is to view change as a continuous process that occurs between any two-time periods. Change is captured by an individual's underlying growth trajectory. Growth models (also known as random coefficients models, multilevel models, and mixed effect models) are well suited for the purpose of studying change at the group as well as the individual level (Bryk and Raudenbush, 1987, Hedeker and Gibbons, 2006, Meredith and Tisak, 1990,

Muthén and Curran, 1997 and Willett and Sayer, 1994). This makes the data analytic approach highly relevant for the analysis of change in randomized experiments. These models have several clear advantages over traditional data analytic approaches for the analysis of repeated-measures data. These advantages include, but are not limited to, the ability to incorporate time-varying predictors, handle dependence among repeated observations in a very flexible manner, and to provide accurate estimates with missing data under fairly unrestrictive missing data assumptions (Duncan and Duncan, 2004, Gueorguieva and Krystal, 2004 and Muthén and Curran, 1997). The data analytic approach also allows researchers to handle other types of dependence among observations due to clustering. For example, effects may, in certain intervention studies, be due to variability among therapists providing the treatment and random effects modeling can be one way to control for clustering due to therapists in experimental designs (Wampold and Serlin, 2000). Finally, although it is largely an unstudied topic, statistical power to detect nonzero parameters can be increased in certain circumstances when using growth models. For example, compared to traditional data analytic approaches (e.g., RM-ANOVA) in the analysis of repeated-measures data from between-group designs, growth models have shown to produce greater statistical power in detecting a difference in a linear slope estimate (i.e., time by group interaction) in small to moderate sample sizes (Fan, 2003).

2.1. Recommendations for design

Although it is possible to apply growth models to two-wave data, the traditional pre–post-treatment design is not optimal, especially if the researcher intends to model individual heterogeneity in growth (Rogosa and Willett, 1985).1 That is, when only two time points are available one cannot separate error variance from individual heterogeneity in change. This precludes the study of individual differences in change and of correlates of individual change. In addition, if the researcher suspects that the functional form of change (time trend) cannot be adequately described with a straight line, two data points are not sufficient. Also, inclusion of multiple measurements during the trial is beneficial when there is missing data. Having more observations on the outcome between pre- and post-treatment assessment can provide valuable information about the individual who fails to return the last assessment point in the study; information that can be used to obtain accurate parameter estimates with incomplete data (I will return to this last point when I discuss missing data in 5.1 and 5.2). Finally, increasing the number of measurements can positively affect the statistical power and reliability of the assessment of individual change (Muthén and Curran, 1997 and Raudenbush and Bryk, 2002).

Thus, more than two time points is often preferred and too few measurements can in certain situations be detrimental to a longitudinal study (Collins, 2006). Several factors can be used to determine the number of time points to be assessed. Consideration needs to be given to the underlying theory of change such as when change is most likely to occur and measure relevant constructs frequently and tightly spaced during these time periods (Collins, 2006). The researcher need also consider the nature of the measures used to assess change, such as whether the measures are sensitive enough to detect a significant change during the time period, whether there is a risk of floor or ceiling effects, and whether repeated testing changes the meaning of the construct being measured (Jackson, 2010). The test–retest reliability of the instrument as well as power issues also factor into decisions about number of measurement occasions. As a general recommendation, researchers should aim to include, at the minimum, four measurement points in a trial because the flexibility in the modeling approach is increased substantially when more than three time points are available. However, whenever possible, the aim should be to include more measurements during the active intervention phase, for example, weekly or biweekly measures of primary outcomes (and process variables) can be beneficial, especially when examining moderators and mediators in a trial.

3. IMPLEMENTING AND INTERPRETING GROWTH MODELS

Growth models can be estimated within a multilevel, mixed model or structural equation modeling framework. Although there are some differences in modeling capabilities, different frameworks often yield identical results (Bollen and Curran, 2006 and Enders, 2011b). In fact, Preacher et al. (2008) point to the fact that differences between frameworks to a large extent are products of the software and as new versions of software are released, differences between frameworks become more difficult to identify. In this paper, I use the generic term growth models (or growth curve models), but readers should be aware that there are some differences between frameworks that will affect results in certain types of models (e.g., models with time-varying predictors). Some models can also be more readily estimated within one framework as compared to another (e.g., parallel process growth models and not missing at random growth models are easier to implement in structural equation modeling framework as compared to linear mixed or multilevel framework). Jackson (2010) has provided some general recommendations when one framework may be preferred over another, but as stated earlier, the software may play an even more important role here.

3.1. An example of a linear growth model in the context of a randomized trial

To illustrate how to implement and interpret a linear growth model, consider the following hypothetical randomized depression trial in which participants were randomly assigned to either an internet-delivered intervention or an active control condition and the primary outcome was measured at four time points weekly throughout the active treatment phase. Fig. 1 illustrates a randomly selected sample of observed individual trajectories and the average mean trend for the entire sample (i.e., the treatment group and control group combined). As can be seen in the figure, individuals started at different levels and changed at different rates over time, but, on average, they seemed to decrease in depression scores over time. Now consider if the researchers wish to fit a linear growth trajectory to the observed data and evaluate whether individuals assigned to the intervention decrease at faster linear rate in depression symptoms as compared to those assigned to the control condition (and hence ending up with lower depression scores at the end of the treatment period).

Growth models can be expressed in the form of a linear regression equation in which the outcome is modeled as function of a temporal predictor variable that captures how the individual changes on the outcome over time. A linear growth model is,

$$Y_{ti} = \beta_0 + \beta_1(TIME_{ti}) + b_{0i} + b_{1i}(TIME_{ti}) + \varepsilon_{ti}$$

where Yti is the outcome at time point t for individual i, TIMEti is the value of the predictor for individual i at time point t (e.g., the time passed since the onset of the study), β0 is the mean intercept, β1 is the mean growth rate (i.e., expected change in the outcome as function of 1 unit change on the time predictor variable), b0i and b1i are so-called random effects that allow the intercepts and growth rates to vary across individuals, and εti is a time-specific residual that expresses the difference between an individual's fitted linear trajectory and the observed data. Thus, the model includes both fixed effects (i.e., averages across individuals) and random effects (i.e., individual deviations from these averages); hence, the use of the term mixed effects models.

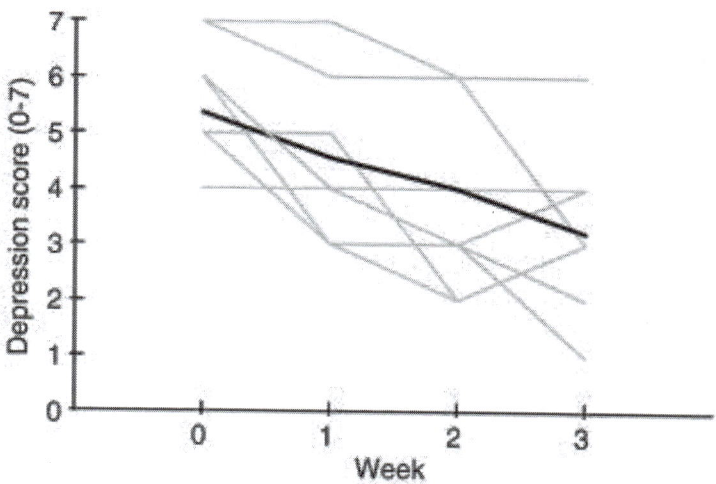

Figure 1. Observed mean trend and randomly selected individual trajectories with artificial depression trial data.

The most noteworthy aspects of the model are the random effects, b0i and b1i, that capture individual heterogeneity (normally assumed to be multivariate normally distributed). Thus, each individual would get his or her own intercept, b0i (here the value on the outcome at initial assessment) and slope, b1i (expected change in the outcome as function of 1 unit change, e.g. one week, on the time predictor variable). In addition, by estimating the covariance between random effects we can obtain information on whether individuals' intercepts and slopes are correlated; that is, in this case, whether there is a relationship between individuals' starting values and individuals' rates of change on the outcome. For example, the researcher may expect that individuals with high initial depression scores will change at faster rate over time (decrease more in depression symptoms) as compared to those with low scores at baseline. This covariance between intercepts and slopes is estimated in the covariance structure associated with random effects (error variance also has an associated covariance structure; I return to this covariance structure in Section 4.1). When only two random effects are included in the model, we can either estimate this covariance (i.e., unstructured covariance structure) or constrain it to be zero (i.e., diagonal covariance structure).

Up until now, the model only examines change over time in the entire sample of individuals and does not reflect the fact the individuals were randomized to different groups. The primary goal of a treatment trial is, of course, to determine whether individuals randomized to the intervention change at a faster (or slower) rate over time as compared to those in the control group. To accomplish this, our linear growth model needs to include a binary coded predictor variable representing conditions (1 = intervention, 0 = control). The model then becomes,

$$Y_{ti} = \beta_0 + \beta_1(TIME_{ti}) + \beta_2(CONDITION_i) + \beta_3(CONDITION_i)(TIME_{ti}) + b_{0i} + b_{1i}(TIME_{ti}) + \varepsilon_{ti}$$

where β0 is the population estimate of the intercept for the control group, β1 is the population estimate of the linear slope for the control group (i.e., control = 0), β2 and β3 capture the estimates of the mean difference in intercept and slope between conditions, b0i and b1i are random effects that allow the intercepts and growth rates to vary across individuals, and εti is a time-specific residual that expresses the difference between an individual's fitted linear trajectory and the observed data. Most relevant, of course, for the primary goal of the study is to test whether β3 (i.e., condition by time interaction; mean difference in slopes between conditions) is nonzero. What is important to note, however, if this coefficient is significantly different from zero it will explain some of the variance in linear slope trajectories, b1i. That is, by including the condition variable in the model we aim to account for the individual heterogeneity and random effects now act as residuals (i.e., unexplained individual heterogeneity not accounted for by condition). In fact, this model is sometimes described as a conditional growth model in which random effects (i.e., intercepts and slopes) are conditioned on the predictor variable, whereas the former model is often referred to as an unconditional growth model because it only includes the temporal predictor of time without other explanatory variables (Bollen and Curran, 2006). Thus, by allowing intercepts and slopes to vary across individuals we can add predictor variables to account for this heterogeneity in growth trajectories.

To aid in the interpretation of this linear growth model, I used artificial data to mimic results of a depression trial and implemented the growth models (unconditional and conditional model) in the linear mixed effects framework using SPSS version 21 (full information maximum likelihood estimation with an unstructured covariance structure for random effects and an identity covariance structure for error variance). The results can be found in Table 1. As can be seen in the table, there were significant individual heterogeneity in initial levels (i.e., intercepts) and slopes. Participants, regardless of assignment, decreased, on average, in depression scores by – 0.71 (β1) points per week. Most noteworthy, the slope heterogeneity could in part be accounted for by the inclusion of the condition predictor variable in the growth model (i.e., the variance associated with random slope was reduced from 0.14 to 0.11, see Table 1). Indeed, there was a significant condition by time interaction, $\beta 3 = -0.40$, indicating that participants assigned to the intervention decreased, on average, by 0.4 points more in depression scores per week as compared to those assigned to control. Thus, condition could systematically explain some of the variance in individual slope estimates. To aid in the interpretation of this effect, Pseudo-R2 can be calculated as the proportion of explained variance in the random effect by the condition predictor variable (Singer and Willett, 2003). By subtracting the estimate of the random effect variance in the conditional growth model (i.e., the model with the predictor) from the estimate of the variance in the unconditional growth model (i.e., the model without the predictor) and by dividing this difference with the unconditional random effects variance, we obtain the proportion of explained variance by the predictor. Based on the information in Table 1, it can be concluded that the condition predictor variable accounted for about 21% of the slope variance ((0.14–0.11) / 0.14 = 21.4%).

3.2. Growth models within structural equation modeling framework

As previously stated, growth models can be implemented in different frameworks. One useful and flexible framework is structural equation modeling. Although a review of structural equation modeling is beyond the scope of this article, here I briefly show how the linear growth model can be cast within this framework. In structural equation modeling, the linear growth model is a two-factor confirmatory factor analysis, where the random intercept and slope are captured by the two latent factors. Fig. 2 depicts a path diagram of the conditional linear growth model. The ellipses are latent variable (with means and variances), rectangles are observed measured variables, single-headed straight arrows represent regression coefficients, and the double-headed arrow denotes covariance. Specifically, the variances associated with the latent variables capture the heterogeneity in growth trajectories (intercept and slopes; b0i, and, b1i) and the mean of the latent intercept and latent slope variable corresponds to the mean of the intercept and slope (β0, and β1; here the average initial level and rate of change in the control condition, as the condition status variable is included in the model). The coefficients for the single-headed arrows that connect the condition status variable (binary coded variable) to the latent variables correspond with the mean differences between conditions (β2 and β3; mean difference in initial level and slope). Finally, the factor loadings for the intercept (1, 1, 1, and 1) and slope (0, 1, 2, and 3) stipulate the functional form of change (i.e., a linear model with equally spaced measurements).

Table 1. Results from a linear growth model with artificial depression trial data.

	Unconditional		Conditional	
	Estimate	SE	Estimate	SE
Fixed effects				
Initial level, β_0	5.33*	0.06	5.39*	0.13
Time, β_1	−0.71*	0.04	−0.39*	0.08
Condition, β_2	–	–	−0.08	0.15
Time by condition, β_3	–	–	−0.40*	0.09
Random effects				
Variance initial level, b_{0i}	0.33*	0.09	0.33*	0.09
Variance slope, b_{1i}	0.14*	0.03	0.11*	0.03
Covariance initial level-slope	0.05	0.04	0.05	0.04
Residual, ε_{ti}	0.69*	0.05	0.69*	0.05

Note. Unconditional growth model does not include condition, whereas the conditional model does. Condition is binary coded variable (1 = intervention, 0 = control). SE = standard error.

* $p < .001$.

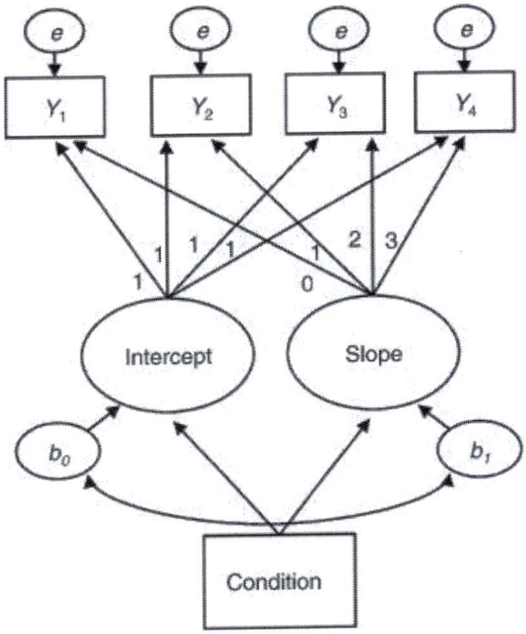

Figure 2. Schematic figure of a linear growth model with a condition status variable (1 = intervention; 0 = control) predicting individual intercepts and slopes.

3.3. Type of predictors in growth models

One of the strengths of this modeling approach is that it can incorporate various forms of predictors in the model. The predictor values may vary over time within individuals, so-called time-varying covariates (e.g., change in medication use over time), or values may be fixed across time within individuals but vary across individuals, so-called time-invariant covariates (e.g., condition in between-group design) (Singer and Willett, 2003). In other words, predictors may either covary with the outcome over time or represent stable characteristics that do not change over time. Time-varying covariates generally account for the variance within individuals (i.e., error variance), whereas time-invariant covariates account for variance between individuals (i.e., random effects; although these variance components are related within a general covariance structure so both within and between variance may be affected with inclusion of both types of predictors, Singer and Willett, 2003 and Wu et al., 2009). An applied example of the use of time-varying covariates can be found in a study that examined associations between the frequency of patient's use of particular words in their text communication with therapists and outcomes in an internet-delivered psychological intervention for generalized anxiety disorder (Dirkse et al., 2014). Researchers can also incorporate regression among random effects in the growth model; in such models, the random effects (intercepts and/or slopes) serve as predictors of other random effects in the model (I return to this type of growth model when I discuss moderation and mediation in 6.1 and 6.2).

3.4. The coding of time and interpretation of the intercept

The interpretation of both the average and individuals' intercepts and slopes changes with the coding of the predictor variables in the equation. Thus, it is necessary to know how time is coded in order to interpret the coefficients. Similar to standard regression analysis, the intercept represents the value at the time point when the predictor variable(s) in the model is zero. For example, if individuals are measured weekly from pre- to post-treatment (4 weeks), a possible coding of time could be 0, 1, 2, and 3. As the first time point is coded as zero, we would interpret the intercept as the average level at the first assessment and individual heterogeneity at that time point (in an unconditional growth model). We could also recode the variable so that the intercept would be at the end of the assessment period, as in, – 3, – 2, – 1, and 0. This would only be a reparameterization of the same model, but it would change the interpretation of the intercept and associated variance (i.e., random effects) (Biesanz et al., 2004). The inclusion of other predictors in the model, besides time, would also change the interpretation of the intercept, so it is import to know how each predictor variable is coded in the model.

Although it is possible to use coding schemes common in standard regression or ANOVA, such as centering or polynomials contrasts, it is generally not recommended (Biesanz et al., 2004). The coding of time should produce readily interpretable estimates that reflect the substantive research question (Biesanz et al., 2004). For example, if researchers wish to evaluate the outcome at the end of the treatment period, making the last measurement point the intercept makes sense. Of course, if one is interested in the time point in the middle of all assessment points (e.g., as a transition point between two phases) centering the time variable is an option (Raudenbush and Bryk, 2002). It is also important to observe that coding of time should reflect the time intervals between measurement occasions; for instance, if measurements were taken every other week, rather than every week, the coding of time would be 0, 2, 4, and 6 (for a linear model).

4. MODEL BUILDING STRATEGIES FOR GROWTH MODELS

In order not to obtain biased parameters estimates, we need to fit an accurate model to the observed data. As with most model building strategies, the goal is to identify a parsimonious statistical model that describes the observed data to satisfactory extent. To identify an appropriate growth model it is important to consider, the time trend, as well as variance within individuals (i.e., error variance) and between individuals (i.e., random effects) (Wu et al., 2009). Given the flexibility of the approach and the many models available to choose from, it is recommended to start the analysis phase by plotting both the average trend and individual trends over time (Kwok et al., 2008). In a randomized controlled trial, plotting these trends as a function of condition can help to identify the functional form of change and allow for the examination of individual heterogeneity. This visual inspection serves to limit the number of models that need to be tested.

4.1. Modifying the time trend and variance components

Several different time trends or functional forms of change can be specified in growth models and this is accomplished by altering how time is coded in the model (Bollen and Curran, 2006). Growth models can incorporate polynomial terms (as standard regression) to capture non-linear change. For example, to obtain a quadratic function we could simply square the linear term (using the aforementioned depression trial example, the quadratic term is 0, 1, 4, and 9) and then include both terms in the model. As before, we can incorporate individual heterogeneity in the trajectories by including random effects associated with the intercept, linear and quadratic terms in the model. As the quadratic function captures the curvature from the straight line, it may be suited for a trial with follow-up data; that is, individuals may change rapidly during the active treatment phase and then change may stabilize during the follow-up phase of the trial. However, an even better data analytic approach in this situation would be to model the treatment trial as qualitatively distinct phases by estimating a separate regression coefficient for each phase of the trial (Duncan and Duncan, 2004 and Muthén and Curran, 1997). This can be accomplished with a piecewise function. With a sufficient number of data points (at least 3 per phase, assuming a linear change model in each phase), researchers may wish to entertain such a model because this would allow them to include all data points in the same analysis and also model change in a way that fits nicely with how change is likely to occur in most trials that include follow-up assessments. In addition, using a piecewise growth model, a set of predictors may be used to predict individual growth in one phase (e.g., pre- to post-treatment) and a separate set may be used during the other phase of the trial (e.g., follow-up phase) (Duncan and Duncan, 2004 and Muthén and Curran, 1997). An example of this model can be found in a study that compared internet-delivered cognitive behavior therapy with face-to-face group therapy for depression over a time period of three years using a randomized design (Andersson et al., 2013).

The choice of functional form of change should be based on theory of change and previous examination of the particular phenomenon under study. Indeed, methodologists have emphasized the importance of theory in model building and design in longitudinal data analysis (Collins, 2006). Sometimes theoretical assumptions about change are nonexistent (Jackson, 2010). In these situations, researchers may choose to test a number of different models to determine the model that best fits the observed data. Several different relative and global fit measures can be used here and some practical guidance for model building is also available (Bollen and Curran, 2006 and Wu et al., 2009).

It is often appropriate to assess the statistical (and practical) significance of model parameters (e.g., means, variance and covariance) when evaluating model fit. Assuming a large enough sample, a common and general way to test the significance of a parameter is the Wald test, in which the point estimate is divided by its standard error; if the ratio exceeds 1.96, it is said to be significantly different from zero at the .05 level. When models are nested, researchers may use a likelihood ratio test to determine whether the inclusion of additional parameters (one or more) in the model contributes significantly to the model (Snijders and Bosker, 2012). The test is a deviance score, which is the difference in – 2 Log Likelihood values between a less restrictive model and a more restrictive one. The significance of the deviance test value is evaluated against the chi-square distribution, where degrees of freedom is equal to the number of parameters that are set to zero in the more restrictive model. Likelihood ratio tests may be useful when examining individual heterogeneity in the form of the inclusion of random effects and associated covariance structure, error variance covariance structures as well as fixed effects (Snijders and Bosker, 2012).2 It should be noted, however, that under certain circumstances, the Wald-test and the likelihood ratio test are too conservative and can produce biased standard errors (which is especially true for variance components as these are bounded; they are by definition nonnegative, making the standard normal symmetric sampling distribution inapplicable) (Snijders and Bosker, 2012). Various corrections and modifications to these commonly used tests can reduce such bias and create more powerful tests for parameters (Molenberghs and Verbeke, 2007). More detailed information on hypothesis tests of fixed and random effects in multilevel models can be found in the text by Snijders and Bosker (2012).

Once the number of random effects has been analytically determined, researchers may also decide to modify the error variance (Kwok et al., 2007). Longitudinal data analysis has historically often incorporated correlated error terms in the model to account for dependence among repeated measurements (e.g., autoregressive structure) (Willett and Sayer, 1994). Dependence is now, at least in part, accounted for by the inclusion of random effects in the model (i.e., dependence associated with that the same individuals are measured repeatedly over time); still, it may be meaningful to test whether the assumption of independence of error variance is tenable. It may also be reasonable to test whether error variance changes over time (i.e., diagonal covariance structure) rather than assuming that variance is homogenous across measurement points (i.e., identity covariance structure) (Willett and Sayer, 1994), which is the default in most multilevel or linear mixed software.

4.2. Overall approach to model building

Unfortunately, the methodological literature offers little clear-cut guidance when it comes to the order in which components should be added to the model and where to start modifying the growth model (Wu et al., 2009). Some recommend estimating all parts in the model and then removing components that do not contribute significantly to the model (e.g., Verbeke and Molenberghs, 2000). Others argue for a bottom-up approach in which one starts with a simple model and evaluates whether the inclusion of additional components adds to the model, working upward from level 1 in multilevel repeated-measures terminology (e.g., Raudenbush and Bryk, 2002 and Snijders and Bosker, 2012).

Wu et al. (2009) provide a baseline model as a starting point for evaluating and re-specifying growth models in both multilevel and structural equation framework. It is often advisable to begin by modifying and evaluating the functional form of change and then adding random effects as needed (see also a recent article that argued that random effects structure should be kept maximal in confirmatory hypothesis testing, Barr et al., 2013). After considering the time trend and taking into account heterogeneity at the individual level, researchers may want to modify the error variance. In a randomized trial, it is often useful to do this process for each condition separately before including all groups in the same model. Finally, predictors (e.g., condition) could be added to account for individual heterogeneity in growth trajectories.

There are no clear and fixed rules to follow when it comes to model specification and both substantive and statistical considerations should be taken into account when choosing among data models (Snijders and Bosker, 2012). As stated earlier, theory should guide researchers and it is often advisable to not test models in isolation, but to compare competing theoretically derived models (Preacher et al., 2008). It is also critical that the specified model accurately reflects the design of the study and the theoretical predictions. This also means that researchers should not drop or include variables in the model solely based on statistical criteria (e.g., significance level). Snijders and Bosker (2012) distinguish between two parts of the data analytic model that can be treated distinctly in the model building phase: the part that involves a priori hypotheses of parameters and the part that is required to get the model to fit the data well and is a prerequisite for valid tests of hypotheses. For the latter part, they argued, an inductive data driven-approach can sometimes be adequate, whereas for the former part of the model a data-driven approach to model specification is not suitable. However, see also Barr et al. (2013) for specific concerns regarding data-driven approaches to the specification of random effect structure in hypothesis testing.

5. STATISTICAL INFERENCES IN GROWTH MODELS WITH MISSING DATA

Missing data is arguably one of the biggest data analytic concerns in a randomized controlled trial. For example, in an intervention study for depression, it is reasonable to expect that participants who do not respond to treatment, and subsequently are the most depressed, are more likely to drop out of the study and not return to complete the follow-up assessment. This is an example of when there is a relationship between outcome and the probability for missing data, so-called non-ignorable missing data (Enders, 2010 and Enders, 2011b). To understand how missing data, and the extent to which missing data, influence statistical inferences, Rubin's (1976) missing data theory is of critical importance. Although the framework has been long established in the methodological literature, it has not always spread to the applied sciences. One reason is probably that the terminology is somewhat confusing and the practical implications of the theory are often elusive to most researchers (Graham, 2009). Three so-called missing data mechanisms are central to the theory: Missing Completely at Random (MCAR), Missing at Random (MAR), and Not Missing at Random (NMAR) (Little and Rubin, 2002). Each one of these mechanisms describes the association between the probability of missing data and other variables and the outcome variable itself. Missing mechanisms are important to consider as they act as assumptions for missing data handling techniques and determine the performance of these techniques (Enders, 2010).

5.1. Ignorable missing data: MCAR and MAR

In the case of MCAR, the probability of missing data on a variable should be unrelated to other measured variables and to the would-be values of that variable (Enders, 2011a). For example, if we have measured participants at four measurement points in a randomized control trial (such as in the depression trial) and we only have missing data at the final time point, MCAR requires that missing, at this time point, is unrelated to treatment group status, to scores on the outcome variable at previous assessments, and to the would-be scores on the outcome variable at the final time point. In other words, if missing data is a random sample of the whole data set then MCAR holds. This is often an unrealistic assumption. Yet, most simple forms of strategies for handling missing data, such as eliminating cases with missing data (e.g., pairwise or list wise deletion) or single regression or mean imputation, assume MCAR. Indeed, numerous simulations studies have shown that these techniques produce biased estimates when data are missing under MAR or NMAR mechanisms (see for reviews, Enders, 2010 and Graham, 2009).

A more lenient missing data assumption is MAR, in which the probability of missing data on a variable is allowed to be related to other observed variables, but not to the would-be values of that variable (Enders, 2011a). Despite what the name implies (random missingness), this is, in fact, systematic missing, because it allows the propensity for missing data to be related to observed variables included in the data analytic model. Graham (2009), for example, proposed that conditional missing at random might be a better word than MAR. Using the aforementioned example, MAR allows the probability of missing data at final assessment to be related to treatment group status, and to scores of previous assessments, but not to the would-be values at the final assessment. Thus, MAR is a much more reasonable assumption than MCAR because it is unlikely that other observed variables are unrelated to the outcome and the propensity for missing data. MAR allows for such associations, assuming that these variables are included in the data analytic model (Enders, 2010).

5.2. Maximum likelihood estimation in growth models

Full information maximum likelihood estimation, as implemented in most software routines that estimate growth models, assumes MAR. Nontechnically, full information maximum likelihood estimation is an iterative process with the aim to identify population parameter values that have the highest probability of reproducing the sample data (readers interested in the technical details of maximum likelihood can consult, Enders, 2010). The technique does not strictly impute missing values, but borrows information from all of the observed data. Using the depression example, individuals who had not completed the final assessment, provided observed data to the estimation process on the first three measurement points and this information was used when estimating all parameters in the model. Assuming a multivariate normal distribution, the information from the incomplete cases steers the estimation process toward more accurate parameter estimates (Enders, 2010 and Enders, 2011a). Thus, rather than disregarding incomplete cases, full information likelihood estimation uses all available data in the estimation process of parameter values. This method of incorporating all individuals in the analysis is most relevant when it comes to a randomized controlled trial for three primary reasons. First, to obtain an unbiased estimate of the average casual effect of the treated requires that all individuals randomized are included in the analysis (i.e., intention-to-treat). Second, if individuals are deleted due to partial missing data, statistical power loss is to be expected. Third, missing data is now handled under MAR rather than MCAR.

Indeed, full information maximum likelihood estimation is one of two recommended methods for handling missing data (the other is multiple imputation) (Schafer and Graham, 2002) and the technique has shown to outperform other methods commonly used in clinical trials. Most noteworthy are direct comparisons between this technique and an ad hoc missing data strategy frequently employed in the clinical trial literature: last observation carried forward. Methodological studies have repeatedly demonstrated that full information maximum likelihood estimation is a better alternative than last observation carried forward because it provides more accurate estimates and standard errors as well as increased statistical power (e.g., Lane, 2008, Mallinckrodt et al., 2001 and Salim et al., 2008). In fact, despite its widespread use, last observation carried forward has shown to produce substantial bias, even under an MCAR mechanism (Enders, 2011a). Thus, full information maximum likelihood estimation has several advantages as compared to traditional remedies for missing data in clinical trials. There are also several examples of studies in the literature of internet interventions that use maximum likelihood estimation to cope efficiently with data loss under MAR (e.g., Hesser et al., 2012, Ljótsson et al., 2014 and Newby et al., 2013).

5.3. Non-ignorable missing data: NMAR

There are situations when MAR-based techniques (e.g., full information maximum likelihood estimation) also yield biased results. NMAR, outcome-dependent missing data or non-ignorable missing, occurs when the probability of missing data on a variable is related to the would-be value of that variable (Enders, 2011b). So returning to our example, NMAR occurs when the likelihood of dropping out of the study before completing the final assessment is directly related to the would-be values of the outcome at that time point of assessment, even after controlling for scores on previous assessments and treatment group status. This is the most problematic situation to handle and unfortunately there is no formal test of MAR (as it is based on an untestable assumption i.e., unobserved data or the would-be values of the outcome), so we cannot determine conclusively whether an NMAR mechanism is at play (Enders, 2011b). Even more problematic is that although several so-called NMAR analyses have been developed over the years (and continue to be developed), these analyses come with their own set of untestable assumptions and even minor violations of these assumptions can result in bias (Enders, 2011b). This has led to methodologists sometimes recommending either that MAR-based techniques be performed even if one suspects NMAR (e.g., Graham,

2009 and Molenberghs et al., 2004), or that a NMAR analysis be conducted as a form of sensitivity analysis to examine whether results hold under various data assumptions and conditions (e.g., Enders, 2011b).

As stated above, statisticians have devoted much energy to develop different forms of NMAR models to handle non-ignorable data (for an overview see, Enders, 2011b), but to date they are not in widespread use in the behavioral sciences. One reason for this is that they have historically been difficult to implement, but new software developments have made these models much more accessible to researchers in applied sciences (Enders, 2011b). One prominent NMAR model for longitudinal data is the pattern mixture model (Hedeker and Gibbons, 1997 and Hedeker and Gibbons, 2006), in which individuals are stratified according to their missing data patterns and a separate growth model is estimated for each pattern (or the patterns are included as dummy predictor variables in the model). The model allows one to examine whether individuals with a certain missing data pattern (e.g., missing data due to attrition) differ in their growth trajectory from those with a complete set of data, and, more importantly, whether these differences affect overall estimates and standard errors. To obtain corrected full sample parameter estimates (e.g., a mean difference in slope between treatment and control), proportion-weighted pattern-specific estimates are averaged across subgroups with different missing data patterns (Enders, 2011b). One can compare this combined overall estimate with the estimate obtained in a traditional MAR-based growth model to determine whether outcome-dependent missing data are likely to have influenced the findings. An applied example of a pattern-mixture model is provided in an open study that examined internet-delivered cognitive behavioral therapy for depression in routine psychiatric care (Hedman et al., 2014; for similar data analytic model in applied setting see also, Hadjistavropoulos et al., 2014).

6. MAKING FULL USE OF GROWTH MODELS: IMPLICATIONS FOR MEDIATION AND MODERATION

Up until now we have considered the strengths of growth models to evaluate the direct effect of intervention on outcome. Yet to realize the full potential and possibilities of these models we should consider how we could apply them to answer substantive research questions related to differences between individuals within intervention studies. Several methodologists have pointed out that these methods respond to central questions in intervention studies, namely who responds best to what intervention and why (e.g., Duncan and Duncan, 2004 and Muthén and Curran, 1997). Questions such as these are commonly answered by the means of statistical analysis of moderation and mediation. A moderator is a variable that affects the strength between the independent (e.g., intervention) and dependent variable (e.g., outcome), whereas a mediator is a variable that is part of casual chain in which the mediator acts as link between the independent variable and dependent variable (Baron and Kenny, 1986).

The importance of the study of moderators and mediators has also been emphasized in the field of internet interventions. For example, Ritterband et al. (2006) made the observation that, "The chief goal of any internet intervention is to produce cognitive and behavior change that leads to symptom improvement. Examining and testing this process using theories and models of behavior change is critical to furthering the understanding of how Internet interventions, and even treatments in general, work." (p. 2). They further argued that, "An advantage of conducting randomized controlled trials through the internet is the ease of obtaining large sample sizes (no geographical limitations), making it possible to better examine mediators and moderators of treatment." (Ritterband et al., 2006, p. 3). To adequately answer questions related to mediation and moderation, researchers need to model change at the individual level. Growth models are therefore of high value when it comes to testing mediators and moderators in intervention studies.

6.1. Examining moderators using growth models

Moderation has often been evaluated using standard regression analysis with an interaction term (MacKinnon, 2008 and MacKinnon et al., 2007). A nonzero interaction effect between moderator and condition in explaining variance in the outcome is interpreted as a moderated treatment effect. That is, the effect of treatment on outcome differs as function of values on the moderator variable. The simplest way to accomplish this with growth models would be to model individual growth on the outcome and then include a potential moderator (continuous or as a dummy variable) as a predictor in the model. A nonzero interaction effect between moderator, group and time would be interpreted as a moderated treatment effect. It is important to remember that by including this new variable in the model we aim to account for individual heterogeneity in slope trajectories. In other words, the question we aim to answer is whether individuals' rate of change varies systematically as a function of both condition and another third variable (i.e., moderator). An applied example of this model can be found in a study that examined moderators in an internet-based prevention trial for eating disorders (Völker et al., 2014). An extension of this moderator growth model would be using the random effect intercept to model the interaction effect between "true" initial status (i.e., intercepts vary over individuals at first assessment point) and condition in their influence on the growth rate. Thus, this model corresponds to a substantive research question, namely whether the intervention effect varies as a function of individuals' initial symptomatology. This can be accomplished by regression among random effects in the structural equation modeling framework (for variations of this model see, Muthén and Curran, 1997).

We can also use more sophisticated approaches to test moderated effects in intervention studies. One potentially useful technique is to empirically classify individuals based on their growth trajectory (intercepts and slopes), so-called group-based trajectory modeling or growth mixture modeling (Nagin and Odgers, 2010). We could add predictors in these types of growth models with the aim of trying to explore relationships between individual baseline characteristics and the empirically based sub-populations of trajectories of change and also examine associations with distal outcomes. Growth mixture models have been specifically developed for the purpose of evaluating intervention effects in sub-populations (Muthen et al., 2002). Although growth mixture models could potentially be very useful when examining moderators in clinical trials, up to date, I am aware of no applied example in the field of internet interventions that has implemented this type of growth model.

6.2. Examining mediators using growth models

In randomized designs, mediation is often evaluated by examining whether the intervention (relative to control) changes the mediator and, if this change, in turn, is correlated with change in the outcome (MacKinnon, 2008). Growth modeling is most relevant when it comes to test mediators in intervention studies. One important reason is that these models allow for the examination of individual differences in change in both the outcome and mediator. Indeed, if one is concerned with how two constructs are related over time, such as a mediator and outcome, the investigation of how change comes about takes precedence. That is, we need an appropriate model for change for both the mediator and the outcome before we can consider if, and how, they are correlated (Cheong et al., 2003). Once an appropriate growth model has been developed for the mediator and outcome, multivariate extensions of growth models then allow researchers to examine whether individual change in the mediator is associated with individual change in the outcome over time in one combined model (Cheong et al., 2003; for an alternative multivariate mediation approach within the multilevel framework see, Bauer et al., 2006). This can be accomplished by regression among random effects in which the random slope of the outcome is regressed on the random slope of the mediator. Support for mediation is found when the treatment group significantly accounts for individual change in the mediator, which, in turn, is associated with individual change in the outcome. Appropriate tests of indirect effects (mediated effects) can then be applied (Cheong et al., 2003 and MacKinnon et al., 2002). A good applied example of this parallel process mediator growth model can be found in a study that compared two forms of internet-delivered psychological treatments for irritable bowel syndrome using weekly measurements of potential mediators and primary outcome (Ljótsson et al., 2013).

One important methodological obstacle for any researcher who aims to evaluate mediation in the context of a randomized controlled trial is the correlational nature of the association between mediator and outcome (MacKinnon, 2008 and MacKinnon et al., 2007). That is, we cannot conclusively determine whether the mediator has a casual effect on the outcome as we do not have experimental control over the mediator. One way to perform a more stringent test of causality in mediator models is to establish the timeline between mediator and outcome; that is, whether change in the mediator preceded and contributed to subsequent change in the outcome (Kazdin, 2007). One way to test this assumption is to include the mediator as lagged time-varying covariate in the growth model. That is, we could predict outcome at time point t using the mediator at time point t – 1 (or any other lag that is relevant) and then reverse the order between the outcome and mediator to compare strength of association. More sophisticated growth models have also been developed with the specific aim to model temporal order between two (or more) variables over time in one combined model, such as the bivariate latent difference scores analysis (McArdle, 2009) and the autoregressive latent trajectory model (Bollen and Curran, 2004).

Finally, growth models can be very useful when researchers test various forms of moderated mediation and mediated moderation models (MacKinnon, 2008 and MacKinnon et al., 2007). An applied example of a moderated mediator model is found in a study that compared two internet-delivered psychological treatments for tinnitus distress (Hesser et al., 2013). Using the mediator as a time-varying covariate in linear mixed effects growth model, the study showed that the treatments produced similar results on the primary outcome, but that different processes in part explained the effects. That is, the mediated effect varied as function of a moderator variable, that is, treatment condition (for technical details see, Bauer et al., 2006). Another interesting model to consider in intervention studies is one in which the mediated effect depends on the baseline level of the mediator (MacKinnon et al., 2007). Similar to the model testing baseline moderated treatment effects, the true initial level on the mediator can be captured by the random intercept in a growth model and the change in mediator and outcome can be modeled with individual growth trajectories (i.e., parallel process growth modeling).

7. REPORTING RESULTS FROM GROWTH MODELS

The overall guiding principle when reporting results is that the analyses should be described in sufficient detail so other researchers are able to replicate them. This guiding principle, of course, also applies to the reporting of results from growth models. Given the flexibility of the approach, this means that researchers need to report more information about the data analytic procedures than they generally do when traditional forms of data models are used (e.g., ANOVA). In other words, it will not suffice to only report, for example, that a linear mixed effects model was conducted to evaluate the outcome of an intervention. Some researchers might find the correct reporting of growth models to be tedious and it is often a tradeoff between clarity, target group and journal page limit/reporting requirements. Still, to report too little information can be a serious problem for the particular study and the field in the long run (Jackson, 2010).

What needs to be reported will to a large extent depend on the growth model that is employed and the particular phenomenon under study. Below, I summarize some general recommendations for reporting of growth models. For more information, Jackson (2010) has provided a guide for the reporting of results from growth models in the multilevel and structural equation modeling framework.

7.1. Recommendations for the reporting of results from growth models

One important aspect to report is the framework used to implement the model, that is, multilevel modeling/linear mixed effects or structural equation framework. As stated earlier, different frameworks often yield identical results; nevertheless it is important to report this so that readers can interpret the findings and replicate the analysis. It is also important to report the software and type of estimation method used. How missing data were handled and under what assumption (MCAR, MAR, or NMAR) should also be communicated clearly in the paper.

Regarding final model(s) interpreted in the results, researchers need to report the functional form of change and describe the coding of time. It is also valuable if researchers provide arguments for these choices in relation to theory of change and/or a substantive research question(s) (Jackson, 2010). The inclusion of random effects (with associated covariance structure) and the choice of error structure (e.g., whether homogenous and independent) are also important to communicate. It is often very useful to include the regression equation for the model(s) in the paper. When estimating models in structural equation modeling framework, researchers may also consider including a schematic figure of the model. Equations and figures can be very informative and be an efficient way to communicate advanced data analytic models (Jackson, 2010).

It is important to describe not only the final model(s), but also the process of model evaluation and re-specification of models during the analysis phase. This includes reporting the procedures and decisions criteria for evaluating and re-specifying models, the type of relative and global fit measures used, and the various models that were tested before selecting the final model(s). It is also important to specify whether models that were tested were determined a priori before the data were collected or were analytically determined post hoc in an exploratory fashion (Jackson, 2010).

In terms of reporting the results of the models, useful summary statistics should be included (e.g., observed and model-implied means, standard deviations, variance). If individual heterogeneity is of particular importance for the study (e.g., in a moderation/mediation study) variance associated with random effects should also be described in the paper, along with Pseudo-R2 (when appropriate) (Singer and Willett, 2003). In such studies, I would also recommend a graphical visualization of individual growth trajectories (Carrig et al., 2004), as this would aid in the interpretation of the results concerning individual differences in change and correlates of individual differences in change. It is often appropriate and useful to report the unstandardized beta coefficients of fixed effects, as this allows readers to interpret the results in the metric of the original outcome scale (e.g., mean difference between groups as function of one unit change in time). Standardized effect sizes for growth models have been developed (Feingold, 2009) and relevant effect sizes (when available) ought to be reported along with test statistics (e.g., z, t or F statistics and degrees of freedom) and confidence intervals for fixed effects. In addition, statistical power to reject a nonzero parameter of primary interest for study (e.g., parameter estimate of group difference at the end of the treatment period) needs to be reported. Power will generally depend on effect size, number of repeated-measurements, sample size (with expected attrition), and the type of data analytic model (e.g., covariance structure, number of random effects, time trend, contrasts) (Hedeker et al., 1999). These aspects, along with alpha level, should be presented when reporting power calculations for growth models.

8. CONCLUDING REMARKS

The primary aim of this paper was to highlight some recent methods in the analysis of change and encourage their use in the field of internet interventions. The particular details of specific advanced models were not given here. There are many excellent books that cover in great detail various forms of growth models, both within the framework of linear mixed effects models/multilevel modeling (Hedeker and Gibbons, 2006 and Singer and Willett, 2003) and structural equation modeling (Bollen and Curran, 2006 and Preacher et al., 2008). Readers who wish to gain a better understanding of these models can do so by consulting these books. In addition, there are specific and important topics that were not fully addressed in this overview. There are several texts that provide in-depth treatments of specific issues, such as estimation and statistical tests of parameters (Snijders and Bosker, 2012 and Verbeke and Molenberghs, 2000), the evaluation of model fit (Wu et al., 2009), the statistical power of rejecting nonzero parameters in between group-designs with attrition (Hedeker et al., 1999), the coding and interpretation of time (Biesanz et al., 2004), and the choice of error structure (Kwok et al., 2007) and random effects structure (Barr et al., 2013).

In this paper, I have argued for the use of growth models in the analysis of change in experiments of internet interventions. These models offer the potential of answering substantive research questions regarding change, individual differences in change and correlates of individual change. The methods also have clear documented advantages for handling common problems in longitudinal data analysis, most significantly, missing data. As such, they have much to offer researchers with respect to treatment evaluation and I believe that these techniques should be seriously considered in studies on internet interventions.

The methods use, however, is also associated with some specific challenges. Growth models require that researchers reconsider the use of the traditional pre–post-treatment design, which is not optimal for these types of models. Given that this general approach to the analysis of change is very flexible, researchers are now also challenged in terms of model building, as they need to choose and argue for an appropriate data analytic model for change. This includes specifying an appropriate time trend and modeling correctly variance within individuals (i.e., error covariance structure) as well as between individuals (i.e., random effects with covariance structure). Finally, these models also require researchers to report sufficient information about models (and model evaluation) so as to be able to replicate the analyses and to interpret the results. Despite these challenges, I do believe that the advantages outweigh the potential drawbacks of using these models in applied research (e.g., underreporting, misspecifying models) and, if implemented correctly, these methods can have an immense impact for the evaluation of direct, indirect and moderated intervention effects in the field of internet interventions.

ACKNOWLEDGMENTS

I would like to thank Brjánn Ljótsson and Heather Hadjistavropoulos who provided valuable comments on an early draft of this article.

REFERENCES

1. Andersson, G., 2009. Using the internet to provide cognitive behaviour therapy. Behav. Res. Ther. 47 (3), 175–180.

2. Andersson, G., Ljotsson, B., Weise, C., 2011. Internet-delivered treatment to promote health. Curr. Opin. Psychiatr. 24 (2), 168–172.

3. Andersson, G., Hesser, H., Veilord, A., Svedling, L., Andersson, F., Sleman, O., Carlbring, P., 2013. Randomised controlled non-inferiority trial with 3-year follow-up of internet-delivered versus face-to-face group cognitive behavioural therapy for depression. J. Affect. Disord. 151 (3), 986–994.

4. Andersson, G., Cuijpers, P., Carlbring, P., Riper, H., Hedman, E., 2014. Guided internet- based vs. face-to-face cognitive behavior therapy for psychiatric and somatic disorders: a systematic review and meta-analysis. World Psychiatry 13 (3), 288–295. Andrews, G., Cuijpers, P., Craske, M.G., McEvoy, P., Titov, N., 2010. Computer therapy for the anxiety and depressive disorders is

effective, acceptable and practical health care: a meta-analysis. PLoS One 5 (10), e13196.

5. Barak, A., Klein, B., Proudfoot, J.G., 2009. Defining internet-supported therapeutic interventions. Ann. Behav. Med. 38 (1), 4–17.

6. Baron, R.M., Kenny, D., 1986. The moderator–mediator variable distinction in social psychological research. Conceptual, strategic, and statistical considerations. J. Pers. Soc. Psychol. 51 (6), 1173–1182.

7. Barr, D.J., Levy, R., Scheepers, C., Tily, H.J., 2013. Random effects structure for confirmatory hypothesis testing: keep it maximal. J. Mem. Lang. 68 (3).

8. Bauer, D.J., Preacher, K.J., Gil, K.M., 2006. Conceptualizing and testing random indirect effects and moderated mediation in multilevel models: new procedures and recom- mendations. Psychol. Methods 11 (2), 142–163.

9. Biesanz, J.C., Deeb-Sossa, N., Papadakis, A.A., Bollen, K.A., Curran, P.J., 2004. The role of coding time in estimating and interpreting growth curve models. Psychol. Methods 9 (1), 30–52.

10. Bollen, K., Curran, P., 2004. Autoregressive latent trajectory (ALT) models: a synthesis of two traditions. Sociol. Methods Res. 32, 336–383.

11. Bollen, K.A., Curran, P.J., 2006. Latent Curve Models: A Structural Equation Perspective. Wiley, Hoboken, NJ.

12. Bryk, A.S., Raudenbush, S.W., 1987. Application of hierarchical linear models to assessing change. Psychol. Bull. 101 (1), 147–158.

13. Carrig, M.M., Wirth, R.J., Curran, P.J., 2004. A SAS macro for estimating and visualizing individual growth curves. Struct. Equ. Model. 11 (1), 132–149.

14. Cheong, J., MacKinnon, D.P., Khoo, S.T., 2003. Investigation of mediational processes using parallel process latent growth curve modeling. Struct. Equ. Model. 10, 238–262.

15. Collins, L.M., 2006. Analysis of longitudinal data: the integration of theoretical model, temporal design, and statistical model. Annu. Rev. Psychol. 57, 505–528.

16. Danaher, B.G., Seeley, J.R., 2009. Methodological issues in research on web-based behav- ioral interventions. Ann. Behav. Med. 38 (1), 28–39.

17. Dirkse, D., Hadjistavropoulos, H.D., Hesser, H., Barak, A., 2014. Linguistic analysis of communication in therapist-assisted internet-delivered cognitive behavior therapy for generalized anxiety disorder. Cogn. Behav. Ther. 1–12.

18. Duncan, T.E., Duncan, S.C., 2004. An introduction to latent growth curve modeling. Behav. Ther. 35 (2), 333–363.

19. Enders, C.K., 2010. Applied Missing Data Analysis. Guilford Press, New York.

20. Enders, C.K., 2011a. Analyzing longitudinal data with missing values. Rehabil. Psychol. 56 (4), 267–288.

21. Enders, C.K., 2011b. Missing not at random models for latent growth curve analyses. Psychol. Methods 16 (1), 1–16.

22. Fan, X., 2003. Power of latent growth modeling for detecting group differences in linear growth trajectory parameters. Struct. Equ. Model. 10 (3), 380–400.

23. Feingold, A., 2009. Effect sizes for growth-modeling analysis for controlled clinical trials in the same metric as for classical analysis. Psychol. Methods 14 (1), 43–53.

24. Francis, D.J., Fletcher, J.M., Stuebing, K.K., Davidson, K.C., Thompson, N.M., 1991. Analysis of change: modeling individual growth. J. Consult. Clin. Psychol. 59 (1), 27–37.

25. Graham, J.W., 2009. Missing data analysis: making it work in the real world. Annu. Rev. Psychol. 60 (1), 549–576.

26. Gueorguieva, R., Krystal, J.H., 2004. Move over ANOVA: progress in analyzing repeated- measures data and its reflection in papers published in the Archives of General Psychiatry. Arch. Gen. Psychiatry 61 (3), 310–317.

27. Hadjistavropoulos, H.D., Pugh, N.E., Nugent, M.N., Hesser, H., Andersson, G., Ivanov, M., Austin, D.W., 2014. Therapist-assisted internet-delivered cognitive behavior therapy for depression and anxiety: translating evidence into clinical practice. J. Anxiety Disord. 28, 884–893.

28. Hedeker, D., Gibbons, R.D., 1997. Application of random-effects pattern-mixture models for missing data in longitudinal studies. Psychol. Methods 2 (1), 64–78.

29. Hedeker, D., Gibbons, R.D., 2006. Longitudinal Data Analysis. Wiley, Hoboken, NJ.

30. Hedeker, D., Gibbons, R.D., Waternaux, C., 1999. Sample size estimation for longitudinal designs with attrition: comparing time-related contrasts between two groups. J. Educ. Behav. Stat. 24 (1), 70–93.

31. Hedman, E., 2014. Therapist guided internet delivered cognitive behavioural therapy. BMJ 348, g1977.

32. Hedman, E., Ljotsson, B., Kaldo, V., Hesser, H., El Alaoui, S., Kraepelien, M., Lindefors, N., 2014. Effectiveness of internet-based cognitive behaviour therapy for depression in routine psychiatric care. J. Affect. Disord. 155, 49–58.

33. Hesser, H., Gustafsson, T., Lunden, C., Henrikson, O., Fattahi, K., Johnsson, E., Andersson, G., 2012. A randomized controlled trial of internet-delivered cognitive behavior therapy and acceptance and commitment therapy in the treatment of tinnitus. J. Consult. Clin. Psychol. 80 (4), 649–661.

34. Hesser, H., Zetterqvist Westin, V., Andersson, G., 2013. Acceptance as a mediator in internet-delivered acceptance and commitment therapy and cognitive behavior therapy for tinnitus. J. Behav. Med. 37 (4), 756–767.

35. Jackson, D.L., 2010. Reporting results of latent growth modeling and multilevel modeling analyses: some recommendations for rehabilitation psychology. Rehabil. Psychol. 55 (3), 272–285.

36. Kazdin, A.E., 2007. Mediators and mechanisms of change in psychotherapy research. Annu. Rev. Clin. Psychol. 3, 1–27.

37. Kwok, O.M., West, S.G., Green, S.B., 2007. The impact of misspecifying the within-subject covariance structure in multiwave longitudinal multilevel models: a Monte Carlo study. Multivar. Behav. Res. 42 (3), 557–592.

38. Kwok, O.M., Underhill, A.T., Berry, J.W., Luo, W., Elliott, T.R., Yoon, M., 2008. Analyzing longitudinal data with multilevel models: an example with individuals living with lower extremity intra-articular fractures. Rehabil. Psychol. 53 (3), 370–386.

39. Lane, P., 2008. Handling drop-out in longitudinal clinical trials: a comparison of the LOCF and MMRM approaches. Pharm. Stat. 7 (2), 93–106.

40. Little, R., Rubin, D., 2002. Statistical Analysis With Missing Data. 2nd ed. John Wiley & Sons, Inc., New Jersey.

41. Ljótsson, B., Hesser, H., Andersson, E., Lindfors, P., Hursti, T., Ruck, C., Hedman, E., 2013. Mechanisms of change in an exposure-based treatment for irritable bowel syndrome. J. Consult. Clin. Psychol. 81 (6), 1113–1126.

42. Ljótsson, B., Hesser, H., Andersson, E., Lackner, J.M., El Alaoui, S., Falk, L., Hedman, E., 2014. Provoking symptoms to relieve symptoms: a randomized controlled dis- mantling study of exposure therapy in irritable bowel syndrome. Behav. Res. Ther. 55, 27–39.

43. MacKinnon, D.P., 2008. Introduction to Statistical Mediation Analysis. Lawrence Erlbaum Associates, New York.

44. MacKinnon, D.P., Lockwood, C.M., Hoffman, J.M., West, S.G., Sheeis, V., 2002. A comparison of methods to test mediation and other intervening variable effects. Psychol. Methods 7 (1), 83–104.

45. MacKinnon, D.P., Fairchild, A.J., Fritz, M.S., 2007. Mediation analysis. Annu. Rev. Psychol. 58, 593–614.

46. Mallinckrodt, C.H., Clark, W.S., David, S.R., 2001. Accounting for dropout bias using mixed- effects models. J. Biopharm. Stat. 11 (1–2), 9–21.

47. Mayo, D.G., 1996. Error and the Growth of Experimental Knowledge. University of Chicago Press, Chicago.

48. McArdle, J.J., 2009. Latent variable modeling of differences and changes with longitudinal data. Annu. Rev. Psychol. 60, 577–605.

49. Meredith, W., Tisak, J., 1990. Latent curve analysis. Psychometrika 55 (1), 107–122.

50. Molenberghs, G., Verbeke, G., 2007. Likelihood ratio, score, and Wald tests in a constrained parameter space. Am. Stat. 61 (1), 22–27.

51. Molenberghs, G., Thijs, H., Jansen, I., Beunckens, C., Kenward, M.G., Mallinckrodt, C., Carroll, R.J., 2004. Analyzing incomplete longitudinal clinical trial data. Biostatistics 5 (3), 445–464.

52. Muthén, B.O., Curran, P.J., 1997. General longitudinal modeling of individual differences in experimental designs: a latent variable framework for analysis and power estimation. Psychol. Methods 2 (4), 371–402.

53. Muthen, B., Brown, C.H., Masyn, K., Jo, B., Khoo, S.T., Yang, C.C., Liao, J., 2002. General growth mixture modeling for randomized preventive interventions. Biostatistics 3 (4), 459–475.

54. Nagin, D.S., Odgers, Candice L., 2010. Group-based trajectory modeling in clinical research. Annu. Rev. Clin. Psychol. 6, 109–138.

55. Newby, J.M., Mackenzie, A., Williams, A.D., McIntyre, K., Watts, S., Wong, N., Andrews, G., 2013. Internet cognitive behavioural therapy for mixed anxiety and depression: a randomized controlled trial and evidence of effectiveness in primary care. Psychol. Med. 43 (12), 2635–2648.

56. Pearl, J., 2009. Causality: Models, Reasoning, and Inference. 2nd ed. Cambridge University Press, New York.

57. Peugh, J.L., Enders, C.K., 2005. Using the SPSS mixed procedure to fit cross-sectional and longitudinal multilevel models. Educ. Psychol. Meas. 65 (5), 811.

58. Preacher, K.J., Wichman, A.L., MacCallum, R.C., Briggs, N.E., 2008. Latent Growth Curve Modeling. Sage, Thousand Oaks, CA.

59. Proudfoot, J., Klein, B., Barak, A., Carlbring, P., Cuijpers, P., Lange, A., Andersson, G., 2011. Establishing guidelines for executing and reporting internet intervention research. Cogn. Behav. Ther. 40 (2), 82–97.

60. Raudenbush, S.W., Bryk, A.S., 2002. Hierarchical Linear Models: Applications and Data Analysis Methods. Sage, Newbury Park, CA.

61. Ritterband, L.M., Tate, D.F., 2009. The science of internet interventions. Introduction. Ann. Behav. Med. 38 (1), 1–3.

62. Ritterband, L.M., Andersson, G., Christensen, H.M., Carlbring, P., Cuijpers, P., 2006. Directions for the International Society for Research on Internet Interventions (ISRII). J. Med. Internet Res. 8 (3), e23.

63. Rogosa, D., 1988. Myths about longitudinal research. Methodol. Issues Aging Res. 171–209.

64. Rogosa, D., Willett, J., 1985. Understanding correlates of change by modeling individual differences in growth. Psychometrika 50 (2), 203–228.

65. Rubin, D., 1974. Estimating causal effects of treatments in randomized and nonrandomized studies. J. Educ. Psychol. 66 (5), 688–701.

66. Rubin, D.B., 1976. Inference and missing data. Biometrika 63 (3), 581–590.

67. Salim, A., Mackinnon, A., Christensen, H., Griffiths, K., 2008. Comparison of data analysis strategies for intent-to-treat analysis in pre-test–post-test designs with substantial dropout rates. Psychiatry Res. 160 (3), 335–345.

68. Schafer, J.L., Graham, J.W., 2002. Missing data: our view of the state of the art. Psychol. Methods 7 (2), 147–177.

69. Singer, J.D., 1998. Using SAS PROC MIXED to fit multilevel models, hierarchical models, and individual growth models. J. Educ. Behav. Stat. 23 (4), 323–355.

70. Singer, J.D., Willett, J.B., 2003. Applied Longitudinal Data Analysis: Modeling Change and Event Occurrence. Oxford University Press, New York.

71. Snijders, T., Bosker, R., 2012. Multilevel Analysis: An Introduction to Basic and Advanced Multilevel Modeling. 2nd ed. Sage, London.

72. Spek, V., Cuijpers, P., Nyklicek, I., Riper, H., Keyzer, J., Pop, V., 2007. Internet-based cognitive behaviour therapy for symptoms of depression and anxiety: a meta- analysis. Psychol. Med. 37 (3), 319–328.

73. Verbeke, G., Molenberghs, G., 2000. Linear Mixed Models for Longitudinal Data. Springer, New York.

74. Völker, U., Jacobi, C., Trockel, M.T., Taylor, C.B., 2014. Moderators and mediators of outcome in Internet-based indicated prevention for eating disorders. Behav. Res. Ther. 63, 114–121.

75. Wampold, B.E., Serlin, R.C., 2000. The consequence of ignoring a nested factor on measures of effect size in analysis of variance. Psychol. Methods 5 (4), 425–433.

76. Willett, J.B., Sayer, A.G., 1994. Using covariance structure analysis to detect correlates and predictors of individual change over time. Psychol. Bull. 116, 363.

77. Wu, W., West, S.G., Taylor, A.B., 2009. Evaluating model fit for growth curve models: integration of fit indices from SEM and MLM frameworks. Psychol. Methods 14 (3), 183–201.

CHAPTER 6

Studies of Climate Change with Statistical-Dynamical Models: A Review

Sergio H. Franchito, Vadlamudi B. Rao

Centro de Previsão de Tempo e Estudos Climáticos, CPTEC, Instituto Nacional de Pesquisas Espaciais, INPE, São José dos Campos, Brazil

ABSTRACT

The cause-effect relationship is not always possible to trace in GCMs because of the simultaneous inclusion of several highly complex physical processes. Furthermore, the inter-GCM differences are large and there is no simple way to reconcile them. So, simple climate models, like statistical- dynamical models (SDMs), appear to be useful in this context. This kind of models is essentially mechanistic, being directed towards understanding the dependence of a particular mechanism on the other parameters of the problem. In this paper, the utility of SDMs for studies of climate change is discussed in some detail. We show that these models are an indispensable part of hierar- chy of climate models.

KEYWORDS

Simple Climate Models, Statistical-Dynamical Models, Climate Change

1. INTRODUCTION

Although the ideal tool for the numerical simulation of climate change would be a fully-resolved three-dimen- sional model, General Circulation Models (GCMs) are not yet able to represent adequately the physical processes of climate and climate change and consequently there are uncertainties for climate change sensitivity. Thus, there is need for the spectrum of models of differing levels of complexity to investigate the climate behaviour, each being optimum for answering specific questions [1] . In this context, simpler climate models are useful to understand large-scale processes and feedbacks acting within the climate system. In addition, simple models are computationally more economical than GCMs and it is relatively easy to analyze and diagnose their behaviour. They can be employed as tools to emulate and understand GCM results.

The simplest climate models are the one-dimensional radiative-convective models (RCMs) [2] . The RCMs usually incorporate relatively detailed treatment of radiative processes and permit for example investigation of the consequences of global scale changes in atmospheric composition. Normally RCMs give only the vertical profile of global mean temperature. Within the class of one-dimension climate models lie the energy balance models (EBMs) wherein vertical integration is done in order to treat latitudinal variation of the climate in terms of surface temperature [3] [4] .

An intermediate class of models are the zonally averaged climate models (ZACMs) in which the treatment of radiative processes can be as sophisticated as in the GCMs but somewhat simplified from the detailed RCMs. While the RCMs and EBMs consider only the thermodynamics of the system, ZACMs include the treatment of dynamics usually by allowing two atmospheric layers. Although the synoptic systems are not included explicitly, their statistical effects are included through parameterization [5] -[11] . Because of this, ZACMs are referred as statistical dynamical climate models (SDMs), although some prefer to make a further distinction [12] . Another advantage of SDMs in comparison with RCMs and EBMs is the inclusion of both latitudinal and vertical variation, thereby permitting more explicit treatment of many feedback mechanisms that must be simplified in the one-dimensional models. While RCMs can treat the vertical structure of the atmosphere, SDMs can additionally estimate the latitudinal characteristics of the response. SDMs are essentially mechanistic, being directed toward understanding the dependence of a particular mechanism on the other parameters of the problem. This is an advantage in comparison to GCMs since the cause-effect relationship is not always possible to trace in these models due to the simultaneous inclusion of several highly complex physical processes. Furthermore, the inter-GCM differences are large and there is no simple way to reconcile them. Since the SDMs are much more computationally efficient than GCMs they can be useful for long-term climate variation studies. In spite of their usefulness they have some restrictions such as the loss of regional resolution. An additional constrain is the errors and limitations that are necessarily introduced in the attempt to parameterize eddy transports based on zonally averaged quantities. Because of their intermediate position, SDMs can be helpful in the design and analysis of GCM studies and in generalizing the results of simple EBM and RCM studies.

Thus, SDMs can be thought of as making a bridge between simpler EBMs and sophisticated GCMs. In a series of papers Franchito and Rao and their collaborators showed the feasibility of using SDMs to study climate change [13] -[22] . In the present work a review of several studies of climate change with SDM is presented. Section 2 shows numerical experiments: Section 2.1 presents the simulation of the annual cycle; Section 2.2 shows the experiments of climate change due land surface alterations; Section 2.3 presents the experiments of climate change due to global warming; and Section 3 presents the latest advances by using statistical-dynamical models; conclusions are shown in Section 4.

2. NUMERICAL EXPERIMENTS WITH SDMS

2.1. Simulation of the Annual Cycle

2.1.1. Simulation of the Annual Cycle of the Atmosphere and Sea Surface Temperature

Since the earth's climate change undergoes significant and gross predictable seasonal changes, a useful test of the validity of a climate model is the ability to simulate the annual cycle. [23] using a two-layer primitive equation SDM simulated the mean annual conditions and zonally averaged atmosphere. In a subsequent paper, [14] applied the model to simulate the sea surface temperature (SST). For this purpose, they assumed that at all latitudes the surface was considered to be covered only by water. The objective was to investigate how well the surface energy balance alone determines SST since no ocean dynamics was included in the model. The results showed that the gross characteristics of the mean annual SST and its seasonal cycle were well simulated by the model. The results suggested that the radiative processes play a fundamental role in determining the SST patterns.

2.1.2. Simulation of the Monsoon-Like Circulations

What characterizes monsoon is the annual variation. The monsoon is associated with a reversal of 180° in the low-level winds from winter to summer. SDMs can effectively be used to study the dynamics of the monsoon. [24] -[27] conducted a series of experiments with ZACMs where the Asiatic continent and adjacent oceans were included. [24] considered a dry monsoon model (i.e., no hydrological cycle) allowing only the effect of a heating differential between an interactive and evolving ocean and the land. The major result was the importance of the east-west ocean land contrast. [25] [26] extended the model scheme to include a full hydrological cycle and compared difference in both the mean seasonal model monsoon and the subseasonal variability of the model monsoon. [27] using the zonally symmetric model of [25] performed a number of experiments in which the physical complexity of the system was successively increased. He found that only with hydrological cycle the low frequency modulations occurred. In these studies the effects of the topography were not taken into account.

[16] used a two-layer primitive equation SDM to verify its ability to capture some monsoon-like variations and to investigate the effects of the topography in the monsoon-like circulation. For this purpose they include into a model a smoothed zonally averaged topography that had a form similar to that observed. The results showed that the model was able to capture some basic characteristics of the monsoon-like circulation such as the seasonal wind reversal (Figure 1(a) & Figure 1(b)). Also, the upper-tropospheric easterly jet in the summer season was well simulated (Figure 1(c) & Figure 1(d)). They

noted that due to the mountain the summer monsoon-like circulation occurred rather suddenly and penetrated farther north. It was also shown that the steepness of the slope (and not the elevation) controls the strength of the monsoon-like circulation in the model.

Other studies have been showed the uselfulness of SDMs to simulate the monsoon circulations not only over Asia but also over West Africa [28] -[31] . The studies regard the effects of vegetation on the monsoon, which will be showed in the next section.

2.2. Climatic Change due to Land Surface Alterations

To our knowledge, very few numerical experiments concerning the climatic effects due to land surface altera- tions have been performed with SDMs. In the experiments of deforestation and desertification realized respec- tively by [32] [33] the changes in geobotanic state were simulated through the modification of the land surface albedo. However, in these experiments changes in geobotanic state were not considered outside the perturbed region. [34] using a version of [35] , incorporated a parameterization of the biofeedback mechanism for the Northern Hemisphere (NH) in which the changes in geobotanic state could also be considered outside the perturbed zone. In this model they considered that a land fraction in each latitude belt was only covered by the predominant type of vegetation. [36] using this SDM, studied the hemispheric response of land surface alterations like deforestation and desertification. The results indicated that the change in evapotranspiration rather than in surface albedo was the predominant effect in regulating the surface temperatures. This is in agreement with that was found by [37] using a GCM.

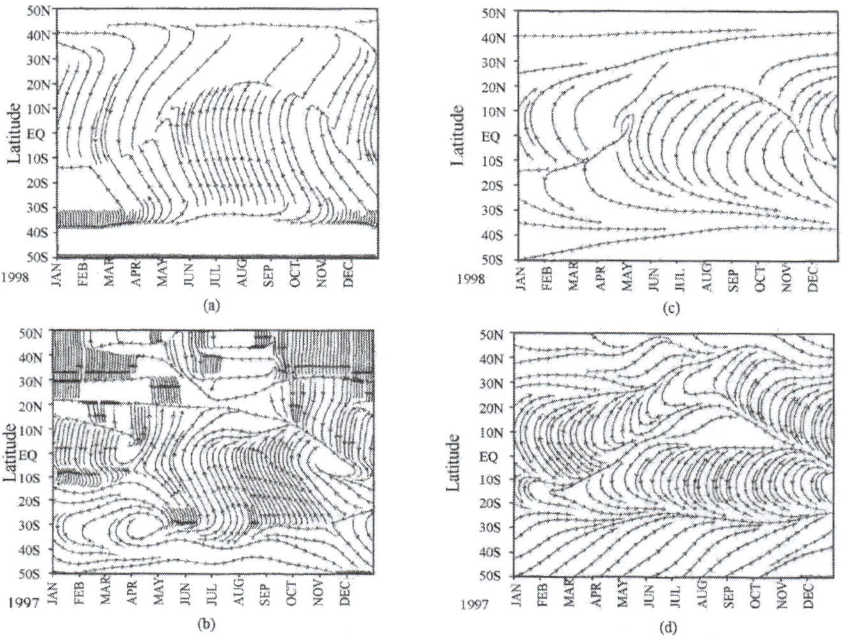

Figure 1. Annual variation of the stream lines: (a) simulated at 925 hPa, (b) observed at 925 hPa and 80°E, (c) simulated at 250 hPa and (d) observed at 250 hPa and 80°E. The observed values are obtained from the NCEP-NCAR reanalysis data. The observations were obtained from the NCEP-NCAR reanalysis data along 80°E, which strongly represent the monsoon. The model winds at 925 hPa were obtained by logarithmic interpolation. (Source: Rao, V. B.; Fernandez, J. P. R.; Franchito, S.H., 2000: Monsoon-like circulations in a zonally-averaged numerical model with topography. Mon. Wea. Rev., 128: 779- 794. ã Copyright [2000] American Meteorological Society. Used with permission).

Although quasi-geostrophic SDMs are adequate for the treatment of the dynamics of the atmosphere in the extratropical region when the interactions between the tropics and higher latitudes are considered the use of the primitive equations is more appropriate. [13]

developed a global primitive equation SDM including a biofeedback mechanism based on the parameterizations of [34] . They showed that the global distribution of the geobotanic zones were well simulated by the model (Figure 2). They applied the model to study the climate effects due to deforestation and desertification. The main results in the two experiments were: in both the hemispheres there was a decrease in the surface net radiation, evapotranspiration and precipitation and an increase of the surface temperature in the perturbed areas showing that the decrease in the evaporative cooling overcomes the effect of the increase of the land surface albedo. These results were similar to those obtained by [36] . However, in [13] the changes were obtained in both the hemispheres.

The treatment of the interaction between the surface processes and the atmosphere is very simple in [13] . [15] incorporated a biosphere model based on BATS [38] in that model. Although well-described complex biosphere models such as BATS have been developed for GCMs, their coupling to simpler SDMs is also relevant for the study of the interactions between vegetation and climate because SDMs greatly simplify analysis and aid the identification of biogeophysical mechanisms. The energy fluxes were computed separately for the land fraction and the remaining part (covered by ocean-ice-snow) of the latitudinal belt. The parameterizations of the biosphere model based on BATS were used for the land fraction of the latitude belt. The biosphere model contained four domains: the subsurface layer, the foliage layer, the air foliage layer and two atmospheric layers (from the surface to 500 hPa and from 500 hPa to the top of the atmosphere). The model involved parameterizations of the energy balance of the Earth's surface, the energy and moisture balances of the foliage air layer and the energy balance of the foliage. The model was applied to study the climate impact due to deforestation and desertification. In the deforestation experiment, the evergreen broadleaf tree in the Amazonian region was substituted by short grass (Figure 3(a)). In the desertification experiment the climatic impact of an anthropogenic degradation of the vegetation situated southward of the Sahara desert was simulated. The land surface modification consisted in the substitution of semi-desert by desert, and tall grass and deciduous shrubs by desert and semi-desert, respectively, in the African continent from 0° to 20°N (Figure 3(b)). The model results were consistent with those obtained from other SDMs, which used parameterization of the biofeedback mechanisms much simpler than BATS [13] [36] . However, in the earlier studies the perturbation was imposed in the entire land fraction of the latitude belt, whereas in [15] the effects of a land surface modification in a determined region of a latitude belt, such as Amazonian deforestation and land degradation southward Sahara desert, can be investigated. The results regarding the changes in the temperature and in the energy fluxes were also in agreement with those of earlier experiments carried out with sophisticated GCMs, which shows the usefulness of this kind of simple model.

Since soil moisture content affects atmospheric conditions by influencing not only the soil albedo, but also the evaporation and hence the energy balance at the surface, it is important to incorporate hydrological processes at the surface. [17] incorporated a soil hydrology model based on BATS and the diurnal cycle in [15] . They investigated the physical feedback of the change in surface characteristics associated with precipitation, evaporation, radiation budget and temperature caused by Amazonian deforestation. They showed that the reduction in transpiration was responsible for the most part of the decrease in total evapotranspiration (63%). The reduction in precipitation was larger than the decrease in evapotranspiration so that runoff was reduced.

Although the SDMs of [15] and [17] were designed to calculate zonal means and not regional features of the Amazonian climate, they allow us to obtain separate simulations for the continental portion of a latitude belt. In South America, most of the continental area of the tropical region is covered by Amazonian forest. Therefore, the effects of Amazonian deforestation on regional climate were analyzed taking into account the model simulations for the land fraction of the tropical region. The simulations must be interpreted as an overall behaviour of climate in the tropical continental region. In general, the changes in temperature and energy fluxes were in good agreement with GCM experiments, showing that the SDMs are able to simulate the characteristics of the tropical climate that are associated with the substitution of forest by pasture areas.

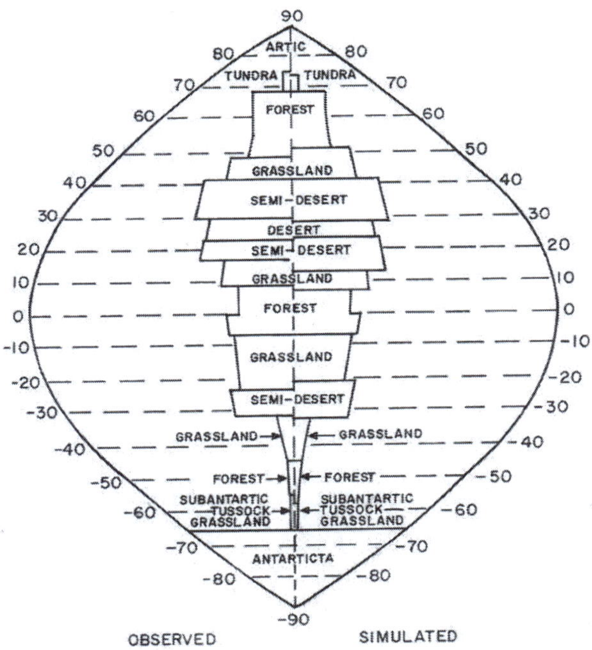

Figure 2. Global distribution of the geobotanic zones simulated by the model (right) and observed (left). (Source: Franchito S. H.; Rao, V. B., 1992: Climatic change due to land surface alterations. Clim. Change, 22, 1-34. ã Copyright [1992] Springer. Used with permission).

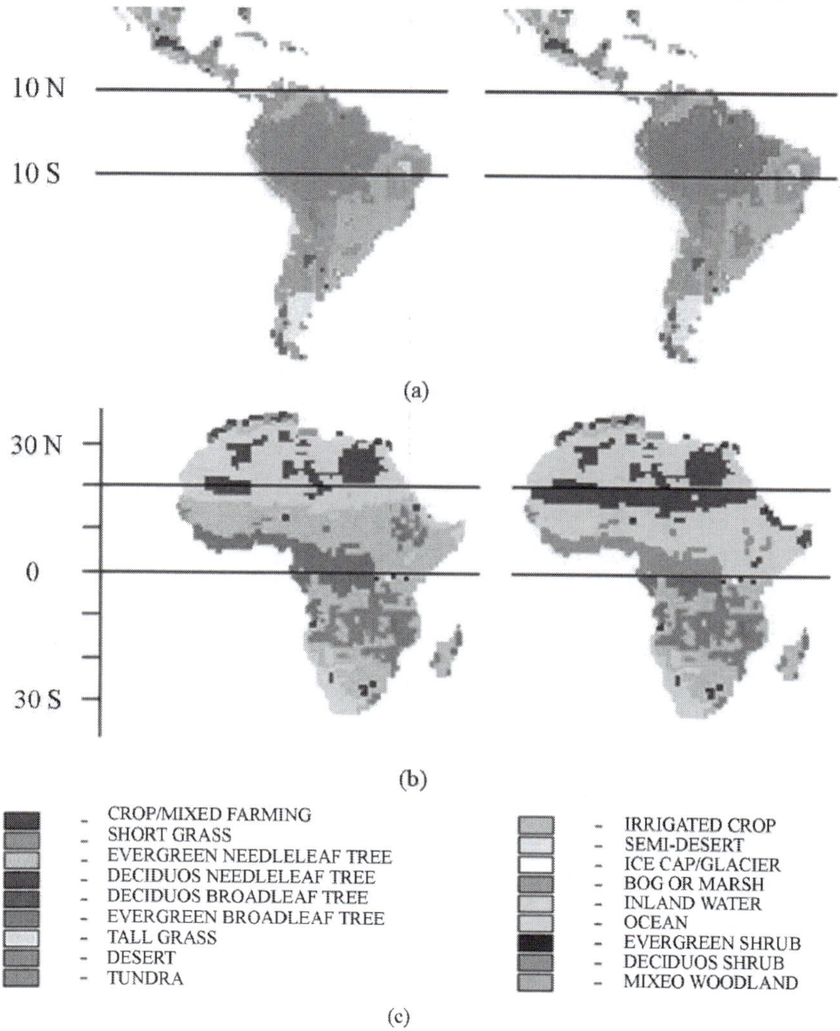

Figure 3. Change in the biomes in: (a) the deforestation experiment and (b) the desertification experiment. (Source: Varejão- Silva, M. A.; Franchito, S. H.; Rao, V. B., 1998: A coupled biosphere-atmosphere climate model suitable for use in climatic studies due to land surface alterations. J. Climate, 11, 1749-1767. ã Copyright [1998] American Meteorological Society. Used with permission).

There have been some studies that investigate the role of change in vegetation on the subtropical Africa. The pioneering work of [39] used an analytical zonally symmetric model to study the qualitative effect of increasing surface albedo on Sahelian rainfall. [40] investigated the impact of sub-Saharan desertification on West African rainfall using a zonally averaged model of the West African monsoon. Their justification for using a two-di- mensional model is based on the fact that West Africa has a zonally uniform distribution of rainfall, vegetation, and other meteorological quantities. [28] reported some preliminary results regarding the relative importance of tropical deforestation and sub-Saharan desertification. In a subsequent paper, [29] presented a more detailed and complete analysis of the problem. They found that changes in vegetation cover along the border between the Sahara desert and West Africa (desertification) may have a minor impact on the simulated monsoon circulation. However, coastal deforestation may cause the collapse of the monsoon circulation and have a dramatic impact on the regional rainfall. [30] developed a zonally symmetric coupled biosphere-atmosphere model including ecosystem dynamics, and applied this model to study biosphere-atmosphere interactions in the region of West Africa. Using this model, they investigated the role of biosphere-atmosphere interactions in the climate variability over West Africa [31] . They demonstrated that the natural response to local grass ecosystem to the dry conditions of the late 1960s played a critical role in maintaining the drought to the following decades in the Sahel region. In a subsequent paper, they suggested that the vegetation dynamics was a significant process in shaping the natural variability of the Sahel rainfall [41] .

2.3. Climate Change Due to Global Warming

Because of significant uncertainty in the behavior of the climate system, evaluations of the impact of an increase in greenhouse gas concentrations in the atmosphere require a large number of long-term climate simulations. In this sense, SDMs can effectively be useful due to their computational efficience. [42] described a 2-D zonally averaged model coupled with a difuse ocean model developed for use in the integrated framework of the Massachusetts Institute of Technology (MIT) Joint Program on the Science and Policy of Global Change (MIT 2-D). The atmospheric model was derived from the Goddard Institute for Space Studies (GISS) Model II GCM [43] and used parameterizations of the eddy transports of momentum, heat and moisture by baroclinic eddies [9] [10] . The results showed that globally averaged values and zonal distributions of equilibrium changes in the different climate variables, such as temperature, precipitation, evaporation, and radiation balance at the surface, as produced by different versions of the 2-D model in response to a doubling of the atmospheric CO_2, were similar to those obtained in simulations with different GCMs. This model was used in various studies regarding the uncertainties of future climate change due to global warming, such as [44] -[47] .

Several studies have examined the impacts of biomass burning in Amazonia on the radiative balance and climate. However, the relative importance of the changes and mechanisms involved has not been investigated. [18] incorporated a detailed radiation model [48] [49] in [15] to study the relative contributions of the changes in the radiation budget and climate caused by smoke aerosols, greenhouse gases and alteration of the land surface characteristics due to biomass burning in the Amazonian forest. To our knowledge, for the first time in that study the effects of the degradation of the surface and smoke aerosols due to biomass burning in Amazonia were investigated together. In general, the greater changes in the radiative balance and climate were due mainly to the changes in the land surface characteristics, followed by those caused by the large amounts of smoke aerosols released in the atmosphere. The changes due to the greenhouse gases were small. The degradation of the surface was responsible for the greatest changes in the net thermal infrared radiation (−14.1%) and net radiation (−17.5%) fluxes at the surface while smoke aerosols seemed to play the main role in controlling the changes in the absorbed solar radiation at the surface (−9.7%). The increase of the air surface temperature was 2°C and 0.7°C in the cases of the degradation of the surface and smoke aerosols, respectively.

[19] used the same model of [18] to evaluate separately each of the four major greenhouse gases (CO_2, O_3, CH_4, and N_2O) in order to quantify their contribution to the future climate change. Such a study had not been done earlier. In the control experiment the actual concentration of the greenhouse gases used in the radiation models was obtained from the IPCC TAR [50] while the concentration of the four major anthropogenic greenhouse gases was those of the more drastic IPCC SRES scenario for 2100 (A1FI-2100). They found that the mean global planetary absorbed solar radiation increased in response to the predicted conditions according to the scenario A1FI for year 2100. This was due to the effect of O_3 absorptions. These increases led to a decrease in the mean global planetary net thermal infrared radiation emitted to space by the earth-atmosphere system to space and to an increase in mean global planetary net radiation. These changes were controlled mainly by the increase in CO_2 concentration (Figure 4). The changes in the radiation budget due to N_2O and CH_4 were small. The change in the air surface temperature response to the predicted conditions for A1FI scenario was mainly controlled by CO_2 concentration (Figure 5).

The future biomes distribution can be modified over the entire globe due to global warming, as projected by IPCC AR4 [51] . Because of the importance of the vegetation-climate interactions in the climate system and their social and economic consequences, more studies using several models of different complexity are needed to improve the knowledge of the impact of global warming on the distribution of the biomes over the globe. [20] used a version of [13] for investigating the impact of the increase of CO_2 concentration on the future global distribution of geobotanic zones. They included a detailed formulation of the radiative transfer models [52] which is suitable for use in ZACMs [53] . The future climate scenarios were obtained from the IPCC AR4 (2007). The results showed that the geobotanic zones over the entire earth can be modified in future due to global warming.

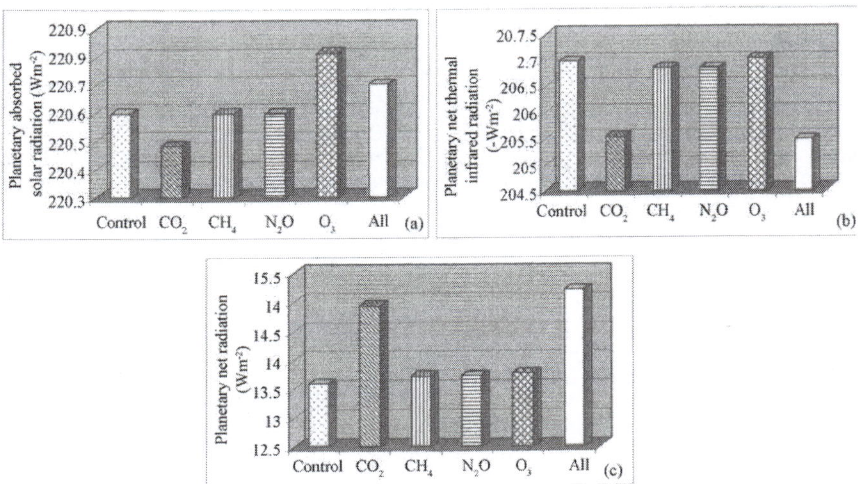

Figure 4. Mean global Planetary (a) absorbed solar radiation, (b) net outgoing thermal infrared radiation, and (c) net radiation at the surface (Wm^{-2}). Shown are the values for the five experiments. (Source: Moraes, E. C.; Franchito, S. H.; Rao, V. B., 2005: Evaluation of surface air temperature change due to the greenhouse gases increase with a statistical-dynamical model. J. Geophys. Res., 110. ã Copyright [2005] American Geophysical Union. Used with permission).

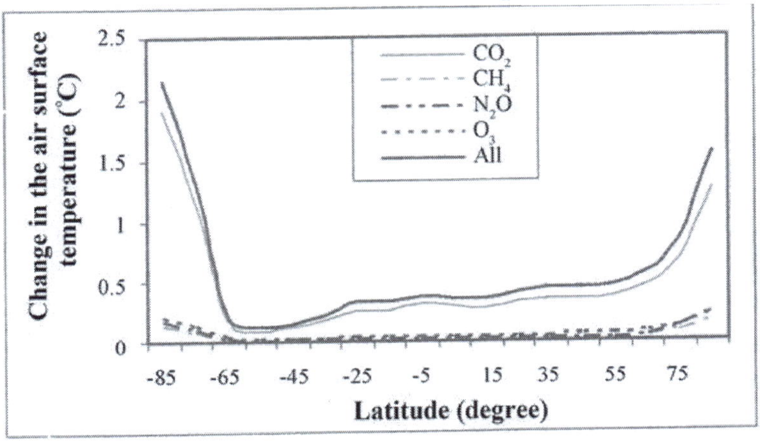

Figure 5. Latitudinal variation of the increase of air surface temperature (°C) (perturbed minus control). Adapted from Moraes et al. (2005) (Sour- ce: Moraes, E. C.; Franchito, S. H.; Rao, V. B., 2005: Evaluation of surface air temperature change due to the greenhouse gases increase with a statistical-dynamical model. J. Geophys. Res., 110. ã Copyright [2005] American Geophysical Union. Used with permission).

Expansion of subtropical desert and semi-desert zones in the NH and SH, retreat of glaciers and sea-ice, with the Arctic region being particularly affected and a reduction of the tropical rainforest and boreal forest can occur due to the increase of the greenhouse gases concentration (see Figure 27 of [20]). The effects were more pronounced in the A1FI and A2 scenarios compared with the B1 scenario. The SDM results confired the IPCC AR4 projections of future climate and were consistent with simulations of more complex GCMs, reinforcing the necessity of the mitigation of climate change associated to global warming.

[21] used the same SDM of [20] to investigate the impact of global warming on the savannization of the tropical land region and the

relative roles of the impact of the increase of greenhouse concentration and future changes in land cover on the regional climate. Their results showed that the climate change due to deforestation was important relative to greenhouse gases at the regional level. The warming due to deforestation corresponded to around 60% of the warming in the tropical region when the increase of CO_2 concentration was included together. However, the global warming due to deforestation was negligible. On the other hand, with the increase of CO_2 concentration projected for 2100 the warming was largely enhanced. The impact of the increase of CO_2 concentration on a deforestation scenario was to increase the reduction of the areas covered by tropical forest (and a corresponding increase in the areas covered by savanna) which may reach 7.5% in future compared with the present climate. Compared with the case with only deforestation, drying may increase by 66.7%. This corroborates with the hypothesis that the process of savannization of the tropical forest can be accelerated in future due to global warming.

More recently, [22] investigated the relative importance of the impact of the land change due to tropical deforestation and global warming on the regional energy balance and climate. The results showed that the higher impact on the energy balance was due to the degradation of land. The percentage of the warming due to deforestation relative to the warming when the increase of greenhouse gas concentration was included together was higher than 60% in the tropical region. This was in agreement with [21] and with other previous studies which suggested that the warming due to deforestation may be important in a regional scale [54] -[56] . On the other hand, with the increase of greenhouse gases concentration an enhancement of the surface temperature occurred. At 5°N the increase relative to deforestation was higher than 50% for the surface temperature and higher than 90% for the foliage and air foliage temperature.

3. LATEST ADVANCES BY USING STATISTICAL-DYNAMICAL MODELS

More recently SDMs makes an important part of Earth System Models of Intermediate Complexity (EMICs). EMICs are complex enough to capture essential climate processes and feedbacks while compromising on the complexity of one or more climate model component. The main characteristic of EMICs is that they describe most of the processes implicit in comprehensive models, albeit in a more reduced (i.e., more parametrized) form. They also explicitly simulate the interactions among several components of the climate system including biogeochemical cycles. On the other hand, EMICs are computationally efficient enough to allow for long-term climate simulations over several tens of thousands of years or a broad range of sensitivity experiments over several millennia. Currently, there are several EMICs in operation such as: two-dimensional, zonally averaged ocean models coupled to a simple atmospheric module [57] [58] or geostrophic two-dimensional [59] or statistical-dynamical [60] atmospheric modules; three-dimensional models with a statistical-dynamical atmospheric and oceanic modules [61] ; reduced-form comprehensive models [62] and those that involve an energy-moisture balance model coupled to an OGCM and a sea-ice model [63] . Results from experiments that were part of the contribution of several EMICs to the 5th Assessment Report of IPCC Working Group 1 are described in [64] [65] .

4. CONCLUSIONS

Manifold physical processes which govern the climate system can be studied using a hierarchy of models. In this paper, a brief discussion of the classification of climate models was given. The advantages of the SDMs were mentioned. Because of their intermediate position in the hierarchy of climate models, SDMs are useful both in the planning of more detailed GCMs and generalizing much simpler 1-D models. For many of the processes where special parameterizations must be developed for a SDM, the questions that force such development are similar to those arising in attempting to treat the subgrid-scale processes needed to improve regional simulations with a GCM. Thus, SDM can and should play an important role in improving understanding of climate system behaviour [66] . Also, SDMs are policy-relevant because they are simpler, faster, and can be used to test specific mechanisms. In this paper, a review of SDMs is presented showing the feasibility of their use to study climate change.

The review presented show that the SDMs are capable of simulating monsoonlike circulations and are useful to study the climate impact due to land surface alterations such as deforestation of tropical Amazonia and desertification of subtropical regions like Sahara as well as the impact of vegetation on the West African monsoon. The SDMs are also useful for studies of the climate effects caused by global warming such as biomass burning in Amazonia and increase of greenhouse gases concentration in future.

REFERENCES

1. Randall, D.A. and Wood, R.A. (2007) Climate Models and Their Evaluation. Climate Change 2007: The Physical Science Basis. Contribution of Working Group I to the Fourth Assessment Report of the Intergovernmental Panel on Climate Change, Cambridge University Press, Cambridge, 590-662.

2. Rasool, S.L. and Schneider, S.H. (1971) Atmospheric Carbon Dioxide and Aerosols: Effects of Large Increases on Global Climate. Science, 173, 138-141.

3. Budyko, M.Y. (1969) The Effect of Solar Radiation Variations on Climate of the Earth. Tellus, 21, 611-619.

4. Sellers, W.D. (1969) A Global Climate Model Based on the Energy Balance of the Earth-Atmosphere System. Journal of Applied Meteorology, 8, 392-400.

5. Saltzman, B. (1968) Steady-State Solutions for the Axially-Symmetric Climate Variables. Pure and Applied Geophysics, 69, 237-259.

6. Saltzman, B. and Vernekar, A.D. (1971) An Equilibrium Solution of the Axially Symmetric Component of the Earth's Macroclimate. Journal of Geophysical Research, 76, 1498-1524.

7. Saltzman, B. and Vernekar, A.D. (1972) Global Equilibrium Solutions for the Zonally Averaged Macroclimate. Journal of Geophysical Research, 77, 3936-3945.

8. Stone, P.H. (1974) The Meridional Variation of the Eddy Heat Fluxes by Baroclinic Waves and Their Parameterizations. Journal of the Atmospheric Sciences, 31, 444-456.

9. Stone, P.H. and Yao, M.S. (1987) Development of a Two-Dimensiooal Zonally Averaged Statistical-Dynamical Model. Part II: The Role of Eddy Momentum Fluxes in the General Circulation and Their Parameterization. Journal of the Atmospheric Sciences, 44, 3796-3786.

10. Stone, P.H. and Yao, M.S. (1990) Development of a Two-Dimensional Zonally Averaged Statistical-Dynamical Model. Part III: The Parametrization of the Eddy Fluxes of Heat and Moisture. Journal of Climate, 3, 726-740.

11. Yao, M.-S. and Stone, P.H. (1987) Development of a Two-Dimensional Zonally Averaged Statistical-Dynamical Model. Part I: The Parameterization of Moist Convection and Its Role in the

General Circulation. Journal of the Atmospheric Sciences, 44, 65-82.

12. Saltzman, B. (1978) A Survey of Statistical-Dynamical Models of Terrestrial Climate. Advances in Geophysics, 20, 183-304.

13. Franchito, S.H. and Rao, V.B. (1992) Climatic Change Due to Land Surface Alterations. Climatic Change, 22, 1-34.

14. Franchito, S.H. and Rao, V.B. (1995) On the Simulation of Sea Surface Temperature with a Zonally Averaged Model. The Global Atmosphere and Ocean System, 3, 3-53.

15. Varejão-Silva, M.A., Franchito, S.H. and Rao, V.B. (1998) A Coupled Biosphere-Atmosphere Climate Model Suitable for Use in Climatic Studies Due to Land Surface Alterations. Journal of Climate, 11, 1749-1767.

16. Rao, V.B., Fernandez, J.P.R. and Franchito, S.H. (2000) Monsoon-Like Circulations in a Zonally-Averaged Numerical Model with Topography. Monthly Weather Review, 128, 779-794.

17. Silva, M.E.S., Franchito, S.H. and Rao, V.B. (2006) Effects of the Amazonian Deforestation on Regional Climate: A Numerical Experiment with a Coupled Biosphere-Atmosphere Model with Soil Hydrology. Theoretical and Applied Climatology, 85, 1-18.

18. Moraes, E.C., Franchito, S.H. and Rao, V.B. (2004) Effects of Biomass Burning in Amazonia on Climate: A Numerical Experiment with a Statistical-Dynamical Model. Journal of Geophysical Research, 109, Article ID: D05109.

19. Moraes, E.C., Franchito, S.H. and Rao, V.B. (2005) Evaluation of Surface Air Temperature Change Due to the Greenhouse Gases Increase with a Statistical-Dynamical Model. Journal of Geophysical Research, 110, Article ID: D24109.

20. Franchito, S.H., Rao, V.B. and Moraes, E.C. (2011) Impact of Global Warming on the Geobotanic Zones: An Experiment with a Statistical-Dynamical Climate Model. Climate Dynamics, 37, 2011-2034.

21. Franchito, S.H., Rao, V.B. and Fernandez, J.P.R. (2012) Tropical Land Savannization: Impact of Global Warming. Theoretical and Applied Climatology, 109, 73-79.

22. Moraes, E.C., Franchito, S.H. and Rao, V.B. (2013) Amazonian Deforestation: Impact of Global Warming on the Energy Balance and Climate. Journal of Applied Meteorology and Climatology, 52, 521-530.

23. Rao, V.B. and Franchito, S.H. (1993) Response of a Simple Model to the Sea Surface Anomalies. Annales Geophysicae, 11, 846-856.

24. Webster, P.J. and Lau, K.M. (1977) A Simple Ocean-Atmosphere Climate Model: Basic Model and a Simple Experiment. Journal of the Atmospheric Sciences, 34, 1063-1084.

25. Webster, P.J. and Chou, L.C. (1980) Seasonal Structure of a Simple Monsoon System. Journal of the Atmospheric Sciences, 37, 354-367.

26. Webster, P.J. and Chou, L.C. (1980) Low Frequency Transition of a Simple Monsoon System. Journal of the Atmospheric Sciences, 37, 368-382.

27. Webster, P.J. (1983) Mechanisms of Monsoon Low-Frequency Variability: Surface Hydrological Effects. Journal of the Atmospheric Sciences, 40, 2110-2124.

28. Zheng, X. and Eltahir, A.E.B. (1997) The Response of Deforestation and Desertification in a Model of West African Monsoons. Geophysical Research Letters, 24, 115-158.

29. Zheng, X. and Eltahir, A.E.B. (1997) The Role of Vegetation on the Dynamics of West African Monsoons. Journal of Climate, 11, 2078-2096.

30. Wang, G. and Eltahir, E.A.B. (2000) Biosphere-Atmosphere Interactions over West Africa. I: Development and Validation of a Couple Dynamical Model. Quarterly Journal of the Royal Meteorological Society, 126, 1239-1260.

31. Wang, G. and Eltahir, E.A.B. (2000) Ecosystem Dynamics and the Sahel Drought. Geophysical Research Letters, 27, 795-798.

32. Potter, G.L., Ellsaesser, H.W., MacCracken, M.C. and Luther, E.M. (1975) Possible Climatic Impact of Tropical Deforestation. Nature, 258, 697-698.

33. Ellsaesser, H.W., MacCracken, M.C., Potter, G.L. and Luther, E.M. (1976) An Additional Model Test of Positive Feedback from High Desert Albedo. Quarterly Journal of the Royal Meteorological Society, 102, 655-666.

34. Gutman, G., Ohring, G. and Joseph, J.H. (1984) Interaction between the Geobotanic State and Climate: A Suggested Approach and a Test with a Zonal Model. Journal of the Atmospheric Sciences, 41, 2663-2678.

35. Ohring, G. and Adler, S. (1978) Some Experiments with a Zonally Averaged Climate Model. Journal of the Atmospheric Sciences, 35, 186-205.

36. Gutman, G. (1984) Numerical Experiments on Land Surface Alterations with a Zonal Model Allowing for Interaction between the Geobotanic State and Climate. Journal of the Atmospheric Sciences, 41, 2679-2685.

37. Nobre, C.A., Sellers, P.J. and Shukla, J. (1991) Amazonian Deforestation and Regional Climate Change. Journal of Climate, 4, 957-988.

38. Zhang, T. (1994) Sensitivity Properties of a Biosphere Model Based on BATS and a Statistical-Dynamical Climate Model. Journal of Climate, 7, 890-913.

39. Charney, J.G. (1975) Dynamics of Desert and Drought in the Sahel. Quarterly Journal of the Royal Meteorological Society, 101, 193-202.

40. Xue, Y., Liou, K.N. and Kasahara, A. (1990) Investigation of Biogeophysical Feedback on the African Climate Using a Two-Dimensional Model. Journal of Climate, 3, 337-352.

41. Wang, G. and Eltahir, E.A.B. (2000) The Role of Vegetation Dynamics in Enhancing the Low-Frequency Variability of the Sahel Rainfall. Water Resources Research, 364, 1013-1021.

42. Solokov, A.P. and Stone, P.H. (1998) A Flexible Climate Model for Use in Integrated Assessments. Climate Dynamics, 14, 291-303.

43. Hansen, J., Russell, G., Rind, D., Stone, P., Lacis, A., Lebedeff, S., Ruedy, R. and Travis, L. (1983) Efficient Three-Dimensional Global Models for Climate Studies: Models I and II. Monthly Weather Review, 111, 609-662.

44. Kamenkovich, L.V., Solokov, A.P. and Stone, P.H. (2002) An Efficient Climate Model with a 3D Ocean and Statistical-Dynamical Atmosphere. Climate Dynamics, 19, 585-598.

45. Forest, C.E., Stone, P.H., Sokolov, A.P., Allen, M.R. and Webster, M. (2002) Quantifying Uncertainties in Climate System Properties with the Use of Recent Climate Observations. Science, 295, 113-117.

46. Sokolov, A.P., Schlosser, C.A., Dutkiewicz, S., Paltsev, S., Kicklighter, D.W., Jacoby, H.D., et al. (2005) The MIT Integrated Global System Model (IGSM) Version 2: Model Description and Baseline Evaluation, MIT JP Report 124.

47. Solokov, A.P., Forest, C.E. and Stone, P.H. (2008) Sensitivity of Climate Change Projections to Uncertainties in the Estimates of Observed Changes in Deep-Ocean Heat Content.

48. Chou, M.-D. and Suarez, M.J. (1994) An Efficient Thermal Infrared Radiation Parameterization for Use in General Circulation Models. Technical Report Series on Global Modeling and Data Assimilation, Technical Memorandum 104606, Goddard Space Flight Center, Greenbelt, 102 p.

49. Chou, M.-D. and Suarez, M.J. (1999) A Solar Radiation Parameterization (CLIRAD-SW) Developed at Goddard Climate and Radiation Branch for Atmospheric Studies. NASA Technical Memorandum NASA/TM-1999-104606.

50. Houghton, J.T., Ding, Y., Griggs, D.J., Noguer, M., van der Linden, P.J., Dai, X., Maskell, K. and Johnson, C.A. (2001) Climate Change 2001: The Scientific Basis. Contribution of Working Group I to the Third Assessment Report of the Intergovernmental Panel on Climate Change, Cambridge University Press, Cambridge, 881 p.

51. Meehl, G.A., Stocker, T.F., Collins, W.D., Friedlingstein, A.T., Gaye, A.T., Gregory, J.M., Kitoh, A., Knutti, R., Murphy, J.M., Noda, A., Raper, S.C.B., Watterson, I.G., Weaver, A.J. and Zhao, Z. (2007) Global Climate Projections. In: Climate Change 2007: The Physical Science Basis. Contribution of Working Group I to the Fourth Assessment Report of the Intergovernmental Panel on Climate Change, Cambridge University Press, Cambridge, 747-845.

52. Jentsch, V. (1991) An Energy Balance Climate Model with Hydrological Cycle. 1. Model Description and Sensitivity to Internal Parameters. Journal of Geophysical Research, 96, 16169-17179.

53. Franchito, S.H., Rao, V.B. and Silva, R.R. (1998) A Parameterization of Radiative Fluxes Suitable for Use in a Statistical-Dynamical Model. Meteorology and Atmospheric Physics, 69, 23-38.

54. Maynard, K. and Royer, J.-F. (2004) Effects of "Realistic" Land-Cover Change on a Greenhouse-Warmed African Climate. Climate Dynamics, 22, 343-358.

55. Feddema, J.J., Oleson, K.W., Bonan, G.B., Mearns, L.O., Buja, L.E., Meehl, G.A. and Washington, W.M. (2005) The Importance of Land-Cover Change in Simulating Future Climates. Science, 310, 1674-1678.

56. Voldoire, A. (2006) Quantifying the Impact of Future Land-Use Changes against Increases in GHG Concentrations. Geophysical Research Letters, 33, Article ID: L04701.

57. Stocker, T., Wright, F.D.G. and Mysak, L.A. (1992) A Zonally Averaged, Coupled Ocean-Atmosphere Model for Paleoclimate Studies. Journal of Climate, 5, 773-797.

58. Marchal, O., Stocker, T.F. and Joos, F. (1998) A Latitude-Depth, Circulation-Biogeochemical Ocean Model for Paleoclimate Studies. Development and Sensitivities. Tellus B, 50, 290-316.

59. Gallée, H., van Ypersele, J.-P., Fichefet, T., Tricot, C. and Berger, A. (1991) Simulation of the Last Glacial Cycle by a Coupled, Sectorially Averaged Climate-Ice Sheet Model I. The Climate Model. Journal of Geophysical Research, 96, 13139-13163.

60. Petoukhov, V., Ganopolski, A., Brovkin, V., Claussen, M., Eliseev, A., Kubatzki, C. and Rahmstorf, S. (2000) CLIMBER-2: A Climate System Model of Intermediate Complexity, Part I: Model Description and Performance for Present Climate. Climate Dynamics, 16, 1-17.

61. Handorf, D., Petoukhov, V.K., Dethloff, K., Eliseev, A.V., Weisheimer, A. and Mokhov, I.I. (1999) Decadal Climate Variability in a Coupled Atmosphere-Ocean Climate Model of Moderate Complexity. Journal of Geophysical Research, 104, 27253-27275.

62. Opsteegh, J., Haarsma, R., Selten, F. and Kattenberg, A. (1998) ECBILT: A Dynamic Alternative to Mixed Boundary Conditions in Ocean Models. Tellus, 50, 348-367.

63. Fanning, A.G. and Weaver, A.J. (1996) An Atmospheric Energy-Moisture Model: Climatology, Interpentadal Climate Change and Coupling to an Ocean General Circulation Model. Journal of Geophysical Research, 101, 15111-15128.

64. Eby, M., Weaver, A.J., Alexander, K., Zickfeld, K., Abe-Ouchi, A., Cimatoribus, A.A., Crespin, E., Drijfhout, S.S., Edwards, N.R., Eliseev, A.V., Feulner, G., Fichefet, T., Forest, C.E., Goosse, H., Holden, P.B., Joos, F., Kawamiya, M., Kicklighter, D., Kienert, H., Matsumoto, K., Mokhov, I.I., Monier, E., Olsen, S.M., Pedersen, J.O.P., Perrette, M., Philippon-Berthier, G., Ridgwell, A., Schlosser, A., Schneider von Deimling, T., Shaffer, G., Smith, R.S., Spahni, R., Sokolov, A.P., Steinacher, M., Tachiiri, K., Tokos, K., Yoshimori, M., Zeng, N. and Zhao, F. (2013) Historical and Idealized Climate Model Experiments: An Intercomparison of Earth System Models of Intermediate Complexity. Climate of the Past, 9, 1111-1140.

65. Zickfeld, K., Eby, M., Weaver, A.J., Alexander, K., Crespin, E., Edwards, N.R., Eliseev, A.V., Feulner, G., Fichefet, T., Forest, C.E., Goosse, H., Holden, P.B., Joos, F., Kawamiya, M., Kicklighter, D.,

Kienert, H., Matsumoto, K., Mokhov, I.I., Monier, E., Olsen, S.M., Pedersen, J.O.P., Perrette, M., Philippon-Berthier, G., Ridgwell, A., Schlosser, A., Schneider von Deimling, T., Shaffer, G., Sokolov, A.P., Spahni, R., Steinacher, M., Tachiiri, K., Tokos, K., Zeng, N., and Zhao, F. (2013) Long-Term Climate Change Commitment and Reversibility: An EMIC Intercomparison. Journal of Climate, 26, 5782-5809.

66. MacCracken, M.C. and Ghan, S.J. (1988) Design and Use of Zonally Averaged Models. In: Schlesinger, M.E., Ed., Physically-Based Modelling and Simulation of Climate and Climate Change, Kluwer Academic Publishers, Dordrecht, 755-809.

CHAPTER 7

Statistical Models for Web Pages Usability

Saad Subair, Hussah AlEisa

College of Computer and Information Sciences, Princess Nourah bint Abdulrahman University, Riyadh, KSA

ABSTRACT

The usability of an interface is a fundamental issue to elucidate. Many researchers argued that many usability results and recommendations lack empirical and experimental data. In this research, the usability of the web pages is evaluated using several carefully selected statistical models. Universities web pages are chosen as subjects for this work for ease of comparison and ease of collecting data. A series of experiments has been conducted to investigate into the usability and design of the universities web pages. Prototype web pages have been developed according to the structured methodologies of web pages design and usability.

Universities web pages were evaluated together with the prototype web pages using a questionnaire which was designed according to the Human Computer Interactions (HCI) heuristics. Nine (users) respondents' variables and 14 web pages variables (items) were studied. Stringent statistical analysis was adopted to extract the required information to form the data acquired, and augmented interpretation of the statistical results was followed. The results showed that the analysis of variance (ANOVA) procedure showed there were significant differences among the universities web pages regarding most of the 23 items studied. Duncan Multiple Range Test (DMRT) showed that the prototype usability performed significantly better regarding most of the items. The correlation analysis showed significant positive and negative correlations between many items. The regression analysis revealed that the most significant factors (items) that contributed to the best model of the universities web pages design and usability were: multimedia in the web pages, the web pages icons (alone) organisation and design, and graphics attractiveness. The results showed some of the limitations of some heuristics used in conventional interface systems design and proposed some additional heuristics in web pages design and usability.

KEYWORDS

Usability, HCI, Web Pages Interface, Statistical Models, Ergonomics

1. INTRODUCTION

User interface is a system layer through which the computer and users communicate [1] - [3] . It is an essential part of human computer interaction (HCI). A survey conducted by Myers and

Rosson [4] showed that software development devoted almost half of the code to the user interface [4] [5] . Dealing with or achieving goals using the interface is known as the usability of the interface. Usability of a system is how easy the system to use and how easy and efficient to perform task [6] -[8] .

The Web is a huge communication system. From a technical point of view, the Web is a system of exchanging data over computer networks using special software. The interfaces used in the web pages are graphical user interfaces that utilize graphics, colours, and icons. Researches showed that there is still a big gap between the researches of the Human Computer Interaction (HCI) and hypertext systems, essentially the web [9] . Shneiderman [6] [10] argued that many researchers' experience lack empirical data to validate or solidify their conclusions. Web sites can be well categorized by the originator's identity such as individual group, university, corporation, and non-profit organization [10] . Information about users can guide web designers to a better design.

The problems that inspired the research of this study can be listed as follows:

1 There were no empirical studies that categorized the users of the universities web pages, studied their preference, and the problems they were facing when using these web pages.
2 To specify the relative importance of some items or parameters of interface design and usability.

The main objective of this study is to investigate into the relations between the items of web pages design and usability and point out which items or variables contribute significantly to web pages design of the universities.

The hypothesis of this study is that:

1 There are significant differences in the design and development of web pages interfaces among the universities;
2 A prototype of a university sample web pages may stimulate and inspire well-structured procedural web pages interface design;
3 Empirical data from a questionnaire and statistical analysis of these data is expected to lead to conclusions and recommendations about the user preference and design aspects of the universities web pages.

The study was focused on the design and development of user interface for the web pages of the universities. The contents of the web pages (the text and the issues inside) were not subjected to detailed studies.

Buckingham and McKnight [9] and Lee and Kozar, [11] reviewed several articles on the web usability and concluded that the link between the hypertext and the web is getting closer. Bieber et al. [12] predicted the evolution of the Web. They examined an array of hypertext technologies which have yet to be adopted in the web development. Smith et al. [13] discussed the information structuring and searching in the Web and reviewed their common problems. Erskine et al. [14] described the application to web site redesign of an adapted form of scenario-based design through claims analysis. Their approaches were contrasting with other theoretical approaches found in the HCI field. They presented a case study of web site redesign which illustrated how scenarios can serve as tools of building interactivity. In a comparative analysis of book-based text and hypertext, Thimble by [15] argued that comprehensive, iterative design and testing (the conventional system development life cycle) of Web structures soon becomes impractical without good authoring tools. He described a tool for Web authoring which uses a page layout language and database of pages to relieve cognitive overhead for authors. Studying distance learning on the web, Benyon et al. [16]

presented a case study of iterative, courseware design. They discussed the different presentation and linking constraints imposed by mark-up tools and browser technology. They concluded that before the web-based learning become a practical and effective option, better authoring tools and clearer pedagogical design principles are needed. The knowledge and experience already gained in other user interface design domains such as menu systems and hypertext [7] [17] - [21] and the results of the research in information retrieval systems [22] [23] can be helpful in web design till empirical data and cases accumulate and give solid and clear methodologies for web pages design. Two of the most important issues in web pages design are length and number of links in web pages. Icon, home, or index page to point to fragments is necessary. Meaningful structures that guide users to the fragments they want is the goal, but excessive fragmentation disrupts those who wish to read or print the full text. Providing users with a visual overview of the web site may reduce the disorientation of the web pages which is a result of adding new materials to the web pages [6] . Staggers [24] suggested that compact vertical design within a page to reduce scrolling of pages can be helpful in web pages design. Some white space can help organize a display of a page. Using excessive horizontal rules or blank lines to separate items is a common mistake in web pages design [25] . Shneiderman [10] suggested that sequencing, clustering, and emphasis for objects within a page are very important issues to consider. The first object in a page is expected to be an important item and likely to be selected (Object/Ac- tion). Clustering related items shows meaningful annotation and relationships between them. Important items can be emphasized with large fonts, color highlights, and surrounding boxes. Questionnaires were used to evaluate the usability of interfaces [26] . However, researchers found problems and weaknesses in the questionnaires developed in the past [10] [27] - [31] . The problems varied from non-represented population sample, to lack of validation and low reliabilities of the questionnaires. Users' acceptance or subjective satisfaction of a system is a critical measure of the system success. Chin et al. [29] developed a measurement instrument which measures the user's subjective rating of the human computer interface called the Questionnaire for User Interface Satisfaction (QUIS). Improvements to the questionnaires for evaluating computer systems were introduced when the sample size and the number of different systems to evaluate were increased [32] . According to the Psychometric Theory the reliability of a questionnaire can be increased by increasing the number of items and scaling steps or points, but much many steps on the scale may affect its reliability adversely (Nunnally [33]).

2. METHODOLOGY

In this study, the problems and the objectives of this research which are stated earlier were attempted to be solved or achieved through:

1 Designing and developing a prototype user interface for a university web page using recent web pages development methodologies.
2 Making the necessary amendments and improvements to the prototype.
3 Evaluating the prototype and other universities web pages interface in a comparative approach through a questionnaire.
4 Using the analysis of variance (ANOVA) procedure to point out the significant differences among the universities web pages with regard to each item (variable) of web pages design.
5 Using correlation analysis to investigate the relationship and association between (items) of universities web pages design variable (each two at a time) as well as multiple relationships
6 Using regression analysis to investigate the causality relationships to find out the most significant independent

variables (items) that form the best model or models for the universities web pages design.

Like other computer programs, user interface systems go through stages or phases of development life cycle. In general, the phases that are usually followed in developing user interface systems are similar to those of con- ventional information system development life cycle with minor modifications. However, Nielsen [2] suggested the usability engineering lifecycle to be particularly applied to interface development and evaluation. The art of prototyping was followed in this study. The first prototype was further improved in sequence till it reached an acceptance level. The first prototype was planned to be a fast design using as little resources as possible. Although HTML is fairly easy programming language, enabling images and adjusting the layouts was somewhat tedious and time consuming. The very first prototype was a scenario where both the different features and functions of the web pages were reduced or cut down. At the next step, this scenario was extended to a horizontal prototype where the surface layer or layout of the web pages (front pages) was in its complete form but the links were not linked to real data or information. In this stage, the concept of prototyping was utilized to the best.

As in conventional system development methodologies, iteration was essential in this design. Most of the steps that were followed needed to be revised. This made the iteration an inevitable process in this design. However, the iteration was done carefully since some iteration might revert the usability of the prototype to some poor usability levels. Shneiderman's Objects/Actions Interface (OAI) Model which follows a hierarchical decomposition of objects and actions in the task and interface domains is considered to decompose complex information problem and building a comprehensible and effective prototype. The benefits from this model were centered in building the required interface objects for the prototype and specifying the required actions on that objects.

2.1. The Experiment

After the prototype was developed, an experiment was conducted to investigate into the usability aspects of four universities web pages and the prototype interface design. The universities are:

1 Univ1;
2 Univ2;
3 Univ3;
4 Univ4;
5 The prototype.

The experiment was a comparative evaluation of the usability problems of these web pages. A sample of 59 students was given a questionnaire to respond to. The students were given enough time to go through and navigate all the five web pages and then asked to attempt answering the questions of the questionnaire. The questions of the questionnaire were mainly designed to point out the usability aspects or attributes of the universities web pages interfaces. The computer laboratory where the experiment was conducted was organized in a way that each user will be independent in his navigation of the web pages and his response to the questions. More than one lab was used for the experiment or evaluation. The evaluation environment was kept to be as normal as possible. Since the test took considerable time (1 to 2 hours), the experiment was conducted in sessions and the users were left free to have a break at any time and respond to the questions in a relax way. The users were informed clearly that all the answers to the questionnaire questions are for the sake of evaluating the web pages and not the users and any preliminary data and personal information will be kept confidential. It is in the guidelines of the questionnaire that, the user may not write his/her name if he/she wishes to.

The answer sheets of the questionnaire were collected after the users finished answering the questions and the data was transferred to the SPSS (Statistical package for social sciences) to be analyzed statistically.

The questionnaire was designed using a rating scale ascending from 1 to 5 which was designed to be administered at controlled experimental conditions. The questions were designed such that each question represents an item of design or heuristic of the interface for the universities web pages. The response of the user (respondent) shows his preference or judgement level of that item. Each item was considered as a variable contributing to the web pages design.

2.2. Items (Variables) Studied

In this study, 23 items or variables were designed as questions in the questionnaire and assigned to the five point scale described before. Nine of the 23 items were concerned with the respondents' knowledge background while the other 14 variables were items of web pages design. The 23 items are listed below.

(a) Respondents Knowledge Background

1 Respondents computer exposure;
2 Respondents internet exposure;
3 Respondents universities web pages exposure;
4 Importance of looking up information on universities in the Internet;
5 Respondents trust of web pages information;
6 Respondents music preference in the universities web pages;
7 Respondents animated pictures preference in the universities web pages;
8 Respondents graphics preference in the universities web pages;
9 Importance of developing web pages for the universities.

(b) Items (Variables) of Web Pages Design

1 Color amount;
2 Contrast against the background;
3 Graphics information provision;
4 Color and Graphics feel of Consistency;
5 Navigation;
6 Location tracking inside the web pages;
7 Hyperlinks (hotspot) semantics;
8 Forms organization and helpfulness;
9 Animated pictures effect on web pages look;
10 Text (alone) organization and design;
11 Icons (alone) organization and design;
12 Icons and text organization and design;
13 Multimedia preference;
14 Universities web pages ranking.

2.3. Statistical Analysis

The data was collected from the questionnaire answer sheets and coded and transferred into SPSSâ statistical package to obtain the statistical analysis for the descriptive statistics of the sample, the analysis of variance (ANOVA) of the universities web pages, and the correlation analysis of the items (variables). The data was then converted into text file and uploaded to the university's mainframe to use SASâ statistical package to obtain the regression analysis output which is clearer in SASâ. Interpretations of all these statistical measures would reveal the significant difference among the universities web pages interface design for each item and point out the better items or variables for the universities web pages design.

2.4. Analysis of Variance Procedure

The ANOVA partitions the source of variations into within groups and between groups variation to obtain the F-value which in turn determines whatever to accept or reject the null hypothesis. The analysis of variance (ANOVA) table will then look as illustrated in Table 1.

Duncan's multiple range test (DMRT) was used to test the differences between groups (universities).

2.5. Correlation Analysis

Simple correlation between the nine respondents' computer and Internet level background items (variable) and the 23 items of universities web pages design (independently) were calculated from the original data using SPSS. The correlation coefficient (r) between two characters (X and Y) is usually expressed as follows:

$$r = \frac{\sum xy}{\sqrt{(\sum x^2)(\sum y^2)}}$$

where

x = x – X and y = y – Y and

x = Independent variable (item); X is the mean of character X.

y = Independent variable (item); Y is the mean of character Y.

2.6. Regression Analysis

Simple and multiple linear regressions of some items (independent variables) on the ranking of the university web pages item (dependent variable) were estimated. The formula for multiple linear regression is as follows:

$$Y = a + b_1X_1 + b_2X_2 + \cdots + b_nX_n$$

where:

Y = dependent variable;

X_{1-n} = independent variables;

a = intercept of the regression line on the Y axis;

b_{1-n} = linear regression coefficients;

n = N Univ3ber of independent variables.

Backward, forward, and stepwise selection regression analysis has also been attempted. In stepwise selection variables are added (or removed) one by one into (from) the model. F-statistic is calculated after a variable is added into (or removed from) the model. A variable that did not produce significant F-value at a redefined probability level is then deleted from (or added to) the model. In forward selection, variables already in the model do not necessary stay there and the process ends when none of the variables outside the model is significantly contributing to the model at the specified probability level [34] [35] .

Table 1. Analysis of variance (ANOVA) table based on individual observations[c].

Source of variation	df[a]	Mean square (variance)	$F_{(Observed)}$[b]
Between groups	k − 1		$Variance_{(Between\ groups)}$
Within groups (error)	N − k		$Variance_{(Within\ groups)}$
Total	N − 1		

[a]k = the n Univ3ber of groups; N = the total n Univ3ber of subjects in the study; [b]Significance level (probability level); [c]Adapted from (Shavelson, 1996).

3. RESULTS AND DISCUSSION

In this article, the nine respondents' knowledge background variables will be discussed first followed by the 14 web pages variables.

3.1. Respondents' Knowledge Background

A correlation analysis study was used to explain the relationship between the nine variables of the respondents' knowledge background. Table 2 shows that respondents' computer exposure has positive significant correlation with the respondents' internet exposure, respondents' exposure to universities web pages, importance of looking up information about universities in the internet, and the importance of developing web pages for universities. However, respondents' computer exposure showed significant negative correlation with the respondents' preference to animation in universities web pages.

Respondents' internet exposure showed positive significant correlation with the respondents' exposure to uni- versities web pages, respondents' trust of the information of the web, and the importance of developing web pages for universities. Meanwhile, respondents' internet exposure showed significant negative correlation with the respondents' music preference and animated pictures preference in the web pages of universities (Table 2). Table 2 also shows that respondents exposure to universities and importance of looking information about universities in the internet followed almost the same pattern of relation with the other variables as the previous variable respondents' internet exposure.

Table 2 also shows that respondents who trust information of the web believe in the importance of developing web pages for universities. It is also shown that respondents who prefer music in the universities web pages they also prefer animated pictures and graphics in the web pages of universities. However, Table 2 shows a negative significant correlation between respondents animation preference in universities web pages and the importance of developing web pages for universities. This result reveals that respondents who believe in the importance of developing web pages for universities do not necessarily prefer animated pictures in these web pages. On the other hand, the table shows that the same respondents prefer graphics in these web pages.

3.2. Items (Variables) of Interface Design and Usability

The descriptive statistics of the 14 items of the web pages interface design and usability are shown in Table 3. The number of observations (N), means, mean standard errors, standard deviations, and variances is shown in Table 3. Low variances in the table show continuous data and hence suggest normal distribution of the data. The analysis of variance (ANOVA) of the variables is shown in Table 4. Only variables that showed significant differences among universities web pages are displayed. Navigation of the web pages, location tracking, hyperlinks semantics, animated pictures effect on the web pages, and icons (alone) organization did not show significant difference in the universities web pages with regards to those variables and hence they are not displayed in the table.

Table 2. Correlation coefficient of the 12 items (variables) of the respondents computers and Internet usage background.

	Respondents computer exposure	Respondents internet exposure	Respondents exposure to universities web pages	Importance of looking information about universities in the internet	Respondents trust of the information of the web	Respondents music preference in the universities web pages	Respondents animation pictures preference in the universities web pages	Respondents graphics preference in the universities web pages
Respondents internet exposure	0.52^{**}							
Respondents exposure to universities web pages	0.40^{**}	0.59^{**}						
Importance of looking information on universities in internet	0.24^{**}	0.13	0.23^{**}					
Respondents trust of the information of the web	0.20^{**}	0.26^{**}	0.18^{*}	0.11				
Respondents music preference in the universities web pages	−0.09	-0.17^{*}	0.02	0.17^{*}	−0.08			
Respondents animation pictures preference in universities web pages	-0.19^{**}	-0.30^{**}	−0.10	-	-0.13	0.55^{**}		
Respondents graphics preference in the universities web pages	0.05	0.05	0.14^{*}	0.07	-	0.36^{**}	0.48^{**}	
Importance of developing web pages for universities	0.33^{**}	0.36^{**}	0.28^{**}	0.38^{**}	0.23^{**}	−0.13	-0.21^{**}	0.20^{**}

However, there are significant differences among the universities web pages regarding the rest of the variables. Duncan's multiple range test (Table 5) was used to arrange these variables in significantly different groups. Table 5 shows that the prototype web pages were grouped in the best group in almost all the variables showed in the table. This reveals that the prototype was almost the best among the other universities web pages.

Table 6 shows the correlation coefficients between the 14 variables studied. The web pages color amount has no significant correlation with any of the variables of the web pages studied. The web pages contrast against the background showed significant positive correlation with the web pages icons and text organization, color and graphics consistency, navigation, location tracking, and forms organization and helpfulness.

However, the contrast against the background showed no significant correlation with the web pages animated pictures effect and icons (alone) organization. The web pages graphics information provision has significant positive correlation with all the other variables studied. Specifically, it showed high positive correlation with the web pages hyperlinks semantics (Table 6). Table 6 shows that web pages color and graphics feel of consistency has positive significant correlation with most of the variables studied. However, it has no significant correlation with the text (alone) and icons (alone) organization but has high positive significant correlation with the icons and text together organization and helpfulness. Navigation of the web pages has significant positive correlation with all the variables studied except the web pages icons (alone) organization and the multimedia preferences in the universities web pages. Location tracking and hyperlinks semantics showed significant positive correlation (separately) with almost all the other variables studied except the ranking of the universities web pages. This result indicated that the ranking of the web pages of the universities did not rely much on these two variables. The web pages forms organization has significant positive correlation with the rest of the variables studied. Forms organization associated significantly with the ranking of the universities web pages. The animated pictures effect on the web pages showed significant positive correlation with the web pages text (alone), icons (alone), and icons and text (together) organization and helpfulness. However, animated pictures have no significant correlation with the ranking of the web pages. Text (alone) organization has significant positive correlation with the icons alone organization and multimedia preferences in the web pages of the universities. However, text (alone) has no significant correlation with the ranking of the universities web pages. Icons (alone) organization has significant positive correlation with the icons and text (together) organization, multimedia preference, and the web pages ranking. Icons and text (together) organization and multimedia preference have significant positive correlation (separately) with the ranking of the universities pages.

Table 3. Number of observations, means, standard errors, standard deviations, and variances of the 14 items (variables) of the universities web pages.

Item (variable)	Number of observation	Mean	Standard error	Standard deviation	Variance
Color amount	292	2.90	0.05	0.78	0.60
Contrast against the background	294	3.29	0.06	1.05	1.10
Graphics information provision	290	3.27	0.05	0.84	0.71
Color and graphics feel of consistency	295	3.12	0.05	0.86	0.73
Navigation	295	3.69	0.05	0.94	0.89
Location tracking inside the web pages	294	3.54	0.06	0.97	0.94
Hyperlinks (hotspot) semantics	295	3.56	0.06	1.01	1.02
Forms organization and helpfulness	290	3.38	0.05	0.87	0.75
Animated pictures effect on web pages look	290	3.61	0.05	0.94	0.88
Text (alone) organization and design	290	3.03	0.06	1.00	1.01
Icons (alone) organization and design	293	3.27	0.06	1.00	0.99
Icons and text organization and design	294	3.62	0.05	0.90	0.80
Multimedia preference	294	3.09	0.06	0.99	0.99
Universities web pages ranking	280	2.97	0.08	1.40	1.96

Table 4. Analysis of variance (ANOVA) tables of the five universities for the 35 items (variables) of the web pages studied[a].

Item (variable)		df	Sum of square	Mean square	F-value	$p > F$
Colour amount						
	Between groups	4	7.422	1.856	3.172	0.014***
	Within groups	287	167.893	0.585		
	Total	291	175.315			
Contrast against the background						
	Between groups	4	18.673	4.668	4.435	0.002***
	Within groups	289	304.171	1.052		
	Total	293	322.844			
Graphics information provision						
	Between groups	4	11.977	2.994	4.431	0.002***
	Within groups	285	192.578	0.676		
	Total	289	204.555			
Colour and graphics feel of consistency						
	Between groups	4	14.590	3.647	5.262	0.000***
	Within groups	290	201.017	0.693		
	Total	294	215.607			
Forms organization and helpfulness						
	Between groups	4	5.876	1.469	1.973	0.099*
	Within groups	285	212.155	0.744		
	Total	289	218.031			
Text (alone) organization and design						
	Between groups	4	9.103	2.276	2.296	0.059*
	Within groups	285	282.552	0.991		
	Total	289	291.655			
Icons and text organization and design						
	Between groups	4	9.526	2.382	3.048	0.017**
	Within groups	289	225.807	0.781		
	Total	293	235.333			
Multimedia preference						
	Between groups	4	10.285	2.571	2.671	0.032**
	Within groups	289	278.236	0.963		
	Total	293	288.520			
Universities web pages ranking						
	Between groups	4	27.086	6.771	3.584	0.007***
	Within groups	275	519.625	1.890		
	Total	279	546.711			

***: Significant at 0.01 level of probability. **: Significant at 0.05 level of probability. *: Significant at 0.10 level of probability. [a]: Only significant variables are displayed.

Table 5. Grouping of the five universities for the 35 items (variables) studied according to Duncan's multiple range test[d].

	Colour amount	Contrast against the back ground	Graphics information provision	Colour and graphics feel of consistency	Navigation	Hyperlink (hotspot) semantics	Forms organization and helpfulness	Text (alone) organization and design	Icons and text organization and design	Multi-media preference	Universities web pages ranking
Univ1	2.74^{a}	3.12^{a}	3.12^{ab}	3.19^{bc}	3.73^{ab}	3.37^{a}	3.38^{ab}	3.19^{b}	3.36^{a}	3.02^{ab}	2.95^{b}
Univ2	2.91^{ab}	3.58^{b}	3.34^{bc}	3.36^{c}	3.78^{ab}	3.59^{ab}	3.40^{ab}	2.95^{ab}	3.80^{b}	3.29^{b}	3.39^{b}
Univ3	3.15^{b}	3.24^{ab}	3.30^{bc}	2.98^{ab}	3.66^{ab}	3.59^{ab}	3.38^{ab}	3.16^{b}	3.58^{ab}	3.20^{b}	3.00^{b}
Univ4	2.73^{a}	2.95^{a}	2.98^{a}	2.76^{a}	3.44^{a}	3.42^{ab}	3.14^{a}	2.72^{a}	3.53^{ab}	2.76^{a}	2.43^{a}
Prototype	2.98^{ab}	3.59^{b}	3.58^{c}	3.32^{c}	3.83^{b}	3.80^{b}	3.59^{b}	3.16^{b}	3.84^{b}	3.19^{b}	3.07^{b}

[d]Means with the same letters indicate no significant difference from each other at the 0.05 probability level as determined by DMRT.

Table 6. Correlation coefficients of the 14 items (variables) of the universities web pages design and usability[a].

	Colour amount	Contrast against the background	Graphics information provision	Colour and graphics feel of consistency	Navigation	Location Tracking	Hypelinks Semantics	Forms organization	Animated pictures effect	Text (alone) organization	Icons (alone) organization	Icons and text organization	Multimedia preference
Contrast against the background	0.05												
Graphics information provision	0.11	0.28**											
Colour and graphics feel of consistency	0.01	0.33**	0.27**										
Navigation	0.09	0.28**	-	0.14*									
Location Tracking	0.05	0.28**	0.25**	0.26**	0.41**								
Hypelinks Semantics	0.03	0.21**	0.34**	0.21**	0.33**	0.40**							
Forms organization	−0.04	0.27**	0.23**	0.30**	0.33**	0.41**	0.37**						
Animated pictures effect	0.00	0.13	-	0.20**	0.10	0.31**	0.22**	0.31**					
Text (alone) organization	0.01	0.20**	0.16**	0.11	0.22**	0.29**	0.26**	0.41**	0.18**				
Icons (alone) organization	0.07	0.09	0.21**	0.11	0.11	-	0.27**	0.21**	0.24**	0.17*			
Icons and text organization	0.06	0.35**	0.28**	0.33**	0.32**	0.44**	0.26**	0.26**	0.35**	-	0.27**		
Multimedia preference	0.13	0.21**	0.20**	0.20**	0.11	0.18**	0.20**	0.29**	-	0.35**	0.24**	-	
Universities web pages ranking	0.11	0.15*	0.15*	0.19**	-	0.03	0.10	0.17*	0.13	0.09	0.18*	0.18**	0.27**

3.3. Regression Analysis

Multiple linear regression results are shown in Table 7. The ANOVA (analysis of variance) table for regression shows that the model was significant at the 0.01 probability level, indicating that the contribution of independent variables coefficients were not equal to zero (alternate hypothesis). The table also shows that multimedia pre- ference in the universities web pages contributed highly and significantly to the model. However, some regression analysis procedures were used, including, forward selection, backward elimination, and stepwise selection. Maximum coefficient of determination (R^2) was achieved by including specific number of independent variables.

Table 7. Linear regression analysis between universities web pages ranking and other items (variables) of web pages design and usability.

Source	Degrees of freedom	Sum of square	Mean square	F	$p > F$
Regression	13	70.00086	5.384682	7.26	0.0001
Error	243	180.1159	0.741218		
Total	256	250.1167			

Variable	Parameter estimate	Standard error	Sum of squares	F	$p > F$
Intercept	1.061532	0.440591	4.302695	5.80	0.0167
Color amount	0.027230	0.075147	0.097328	0.13	0.7174
Contrast against the background	−0.030090	0.059111	0.192028	0.26	0.6112
Graphics information provision	0.068228	0.074656	0.619078	0.84	0.3617
Color and graphics feel of consistency	−0.030560	0.071732	0.134551	0.18	0.6704
General navigation	0.069340	0.068237	0.765360	1.03	0.3106
Location tracking	−0.064310	0.069323	0.637798	0.86	0.3545
Hyperlinks (hotspots) semantics	0.121541	0.073557	2.023700	2.73	0.0998
Forms organization and helpfulness	0.026652	0.070631	0.105537	0.14	0.7063
Animated pictures	−0.055670	0.089501	0.286755	0.39	0.5345
Text (alone) organization and design	−0.109380	0.064121	2.157001	2.91	0.0893
Icons (alone) organization and design	0.200612	0.057933	8.888093	11.99	0.0006
Icons and text organization and design	−0.018650	0.060757	0.069868	0.09	0.7591
Multimedia preference	0.416697	0.073722	23.68064	31.95	0.0001

Table 8 shows a summary of forward selection procedure for the dependent variable universities web pages ranking. The procedure showed that only eight independent variables contributed significantly to the model. However, universities web pages multimedia preference and icons (alone) organization and design were contributed significantly to the model. The forward selection begins with the most independent variable contributing highly to the model and advances to the next most contributing one till it reaches a satisfactory model.

Unlike the forward selection, backward elimination begins with the complete model including all the independent variables and then started to eliminate the most non-significant variable in the model considering its F-value, then proceed to the next most non-significant one. Table 9 shows a summary of the backward elimination procedure. As seen in the table, nine variables were removed from the model. This indicated that these 9 variables did not contribute significantly to the model (in this order) as the rest of the variables did.

The forward selection and backward elimination adds and eliminates (respectively) independent variables to (from) the model regardless of what variables are already outside or existing in the model. Stepwise selection re- examines at every stage the regression of the variables incorporated into the model in previous stages. A variable which might have been the best single variable to enter the model at an early stage, might at later stage, be superfluous because of the relationships between it and other variables now in the regression model [34] . Table 10 shows a summary of stepwise selection regression procedure model which included only four variables that were significantly contributed to the model at the 0.15 probability level. That is to say, universities web pages multimedia preference, icons (alone) organization, text (alone) organization, and hyperlinks (hotspots) semantics were the four most independent variables contributing to the model significantly.

Coefficient of determination (R^2) measures the proportion of total sum of squares of the variables that is explained by the regression line. It is a measure of how closely the points (observation) fit the least square line. Consequently, the line that has maximum R^2 represents the best fitting line [36] . Table 11 shows the maxi mum R^2 that is attained with one independent variable, two independent variables, and so on till 13 independent variables. This might indicate the ranking importance of an independent variable as a single or in combination with other independent variables to universities web pages ranking. The results showed that the universities web pages location tracking, hyperlinks semantics, text (alone) organization, icons (alone) organization, and design, and multimedia preferences were among the first five variables (items) that contributed significantly to the universities web pages ranking. However, limitation to R^2 is that the addition of an irrelevant independent variable to the regression model will increase R^2 even though the irrelevant variable is not related to the other variables in the model [36] . The assumption of this model is that all the variables (items) have been estimated precisely.

Table 8. Summary of forward selection procedure for dependent variable universities web pages ranking.

Step	Variable entered	Number in	Partial R^2	$P > F$
1	Multimedia preference	1	0.2132	0.0001
2	Icons (alone) organization and design	2	0.0401	0.0003
3	Text (alone) organization and design	3	0.0073	0.1151
4	Hyperlinks Semantics	4	0.0094	0.0736
5	Location Tracking	5	0.0021	0.3959
6	Navigation	6	0.0019	0.4177
7	Graphics information provision	7	0.0017	0.4511
8	Animated pictures effect	8	0.0017	0.4466

Table 9. Summary of backward elimination procedure for dependent variable universities web pages ranking.

Step	Variable removed	Number in	Partial R^2	$P>F$
1	Icons and text organization	12	0.0003	0.7591
2	Color amount	11	0.0004	0.7161
3	Forms organization	10	0.0003	0.7426
4	Color and graphics feel of consistency	9	0.0006	0.6585
5	Contrast against the background	8	0.0009	0.5751
6	Animated pictures effect	7	0.0017	0.4466
7	Graphics information provision	6	0.0017	0.4511
8	Navigation	5	0.0019	0.4177
9	Location tracking	4	0.0021	0.3959

Table 10. Summary of stepwise procedure for dependent variable universities web pages ranking[a].

Step	Variable entered/removed	Number in	Partial R^2	Model R^2	$P>F$
1	Multimedia preference	1	0.2132	0.2132	0.0001
2	Icons (alone) organization and design	2	0.0401	0.2534	0.0003
3	Text (alone) organization and design	3	0.0073	0.2607	0.1151
4	Hyperlinks semantic	4	0.0094	0.2700	0.0736

[a]: All variables in the model are significant at 0.15 probability level.

The results of Tables 7-11 suggest that, some independent variables significantly contributed to the regression model and others did not. For a single independent variable, multimedia preference of the universities web pages gave the most significant contribution to the dependent variable universities web pages ranking (Table 11). For two independent variables, icons (alone) organization and design and multimedia preference of the universities web pages gave the most significant contribution to the dependent variable universities web pages ranking. For three independent variables, the universities web pages text (alone) organization and design, icons (alone) organization and design, and multimedia preference gave the most significant contribution to the dependent variable universities web pages ranking (Table 11).

Table 11. R^2 of single and/or combination of different independent variable (s) with dependent variable universities web pages ranking.

Number of variables in the model	Independent variable (s)	R^2
1	Multimedia preference	0.2132491
1	Icons (alone) organization	0.084938
1	Hyperlinks semantics	0.055802
2	Icons (alone) organization and design and multimedia preference	0.253370
2	Hyperlinks semantics and multimedia preference	0.222795
2	Navigation and multimedia preference	0.218554
3	Text (alone) organization and design and icons (alone) organization and design and multimedia preference	0.260674
3	Hyperlinks semantics and icons (alone) organization and design and multimedia preference	0.259198
3	Color amount and icons (alone) organization and design and multimedia preference	0.256156
4	Hyperlinks semantics and text (alone) organization pages and icons (alone) organization and design and multimedia preference	0.270027
4	Navigation and text (alone) organization pages and icons (alone) organization and design and multimedia preference	0.263917
4	Graphics information provision and text (alone) organization pages and icons (alone) organization and design and multimedia preference	0.263123
5	Location tracking and hyperlinks semantics and text (alone) organization and icons (alone) organization and design and multimedia preference	0.272125
5	Graphics information provision and hyperlinks semantics and text (alone) organization pages and icons (alone) organization and design and multimedia preference	0.271217
5	Navigation and hyperlinks semantics and text (alone) organization pages and icons (alone) organization and design and multimedia preference	0.271186

4. CONCLUSIONS

From the discussion, some important conclusions and findings concerning both users (respondents) and items (variables) of the web pages of the universities [37] were unveiled. These conclusions and findings can be summarized as follows:

1 Respondents who were frequent in using computers did not prefer animated pictures in universities web pages (Table 2).
2 Respondents who frequently access the Internet believe that it is important to develop web pages for universities (Table 2).
3 Respondents who were well exposed to computer did not prefer music and animation in the universities web pages (Table 2).
4 There were significant differences between the universities web pages regarding most of the variables studied (Table 4).
5 The prototype which was developed according to a scientific methodology was proved to be better than most of the other universities web pages (Table 5).
6 Good contrast against the background of the web pages made the navigation of the web pages more convenient (Table 6).
7 Good design of graphics made the icons and hyperlinks meaningful and had metaphoric dimensions (Table 6).
8 The usage of the icons and text together in the universities web pages design was proved to be better than the usage of the text alone or the icons alone (Table 6).
9 Multimedia in the web pages, the web pages, icons organization and design, the web pages text organization and design, and hyperlinks semantics in the web pages independently and separately contributed significantly (in this order) to the ranking of the universities web pages (Table 7).
10 The best five items of web pages design and usability that together formed the most significant model were: location tracking, hyperlinks semantics, text (alone) organization and design, icons (alone) organization and design, and multimedia preference (Table 11).

5. RECOMMENDATIONS FOR FUTURE WORK

1 Further studies can be done to increase the validity and reliability of the questionnaire used in this study. This can be

achieved by increasing and randomizing the sample of the test users (respondents), increasing the steps of the scale (e.g. 10 points scale), and increasing the number of the questions used in the questionnaire.

2 In further studies, it is suggested that the questions of the questionnaire can be improved by being more specific and adhere to the scale designed for that. General questions can yield confounding results. For one item or variable of interface usability there may be several questions instead of only one or few.

3 In this study, only one sample group was used. In future work, the sample users (respondents) can be divided into two groups: users and experts. More sophisticated scale (e.g. 10 points scale) can be constructed for expert users (respondents). In 10 points scale for instance, expert users can better distinguish the different levels (points) of a single question which can yield a wide range of data which could be better analyzed statistically.

4 Additional statistical models can rigorously analyze the usability of the web pages and hence more valid and verified conclusions can be achieved.

ACKNOWLEDGEMENTS

The authors would like to thank Princess Nourah bint Abdulrahman University Riyadh, KSA, for assistance they rendered concerning this work.

REFERENCES

1. Subair, S. (2014) The Usability Engineering of the Universities Web Pages User Interfaces: A Pragmatic Approach. Princess

Nourah bint Abdulrahamn University Printing Press, Riyadh, in Press.

2. Retherfor, R.D. and Choe, M.K. (1993) Statistical Models for Causal Analysis. John Wiley & Sons, Inc., New York.
3. SAS Institute Inc. (1987) SAS /STATTM Guide for Personal Computers. Version 6, SAS Institute Inc., Cary.
4. Drapper, N.R. and Smith, H. (1981) Applied Regression Analysis. John Wiley & Sons Inc., New York.
5. Nunnally, J.C. (1978) Psychometric Theory. McGraw-Hill Book Company, New York.
6. Rushinek, A. and Rushinek, S.F. (1986) What Makes User Happy? Communication of the ACM, 29, 594-598.
7. Larcker, D.F. and Lessig, V.P. (1980) Perceived Usefulness of Information: A Psychometric Examination. Decision Science, 11, 121-134.
8. Ives, B., Olson, M.H. and Baroudi, J.J. (1983) The Measurement of User Information Satisfaction. Communication of the ACM, 26, 785-793.
9. Chin, J.P., Virginia, A.D. and Norman, K.L. (1988) Development of an Instrument Measuring User Satisfaction of the Human-Computer Interface. Proceedings of the SIGCHI Conference on Human Factors in Computing Systems, Paris, 27 April-2 May 2013, 213-218.
10. Chin, J.P., Norman, K.L. and Shneiderman, B. (1987) Subjective User Evaluation of CF PASCAL Programming Tools. Technical Report (CAR-TR-304), Human-Computer Interaction Laboratory, University of Maryland, College Park.

11. Bailey, J.E. and Person, S.W. (1983) Development of a Tool for Measuring and Analysing Computer User Satisfaction. Management Science, 29, 530-545.

12. Norman, D. and Nielsen, J. (2013) 10 Usability Heuristics for User Interface Design. http://www.nngroup.com/articles/ten-usability-heuristics/

13. Horton, W., Taylor, L., Ignacio, A. and Hoft, N.L. (1996) The Web Page Design Cookbook. John Wiley & Sons, Inc., New York.

14. Staggers, N. (1993) Impact of Screen Density on Clinical Nurses' Computer Task Performance and Subjective Screen Satisfaction. International Journal of Man-Machine Studies, 39, 775-792.

15. Marchionini, G. (1989) Making the Transition from Print to Electronic Encyclopaedia: Adaptation of Mental Models. International Journal of Man-Machine Studies, 30, 591-618.

16. Belkin, N.J. and Croft, B.W. (1992) Information Filtering and Information Retrieval: Two Sides of the Same Coin? Communications of the ACM, 35, 29-38.

17. Isakowitz, T., Stohr, E.A. and Balasubramanian, P. (1995) RMM: A Methodology for Hypermedia Design. Communications of the ACM, 38, 34-44.

18. Rivlin, E., Rotafogo, R. and Shneiderman, B. (1994) Navigating in Hyperspace: Designs for a Structure-Based Toolbox. Communications of the ACM, 37, 87-96.

19. Norman, K. (1991) The Psychology of Menu Selection: Designing Cognitive Control at the Human/Computer Interface. Ablex, Norwood.

20. Shneiderman, B. and Kearsley, G. (1989) Hypertext Hands-On: An Introduction to a New Way of Organising and Accessing Information. Addison-Wesley, Reading.

21. Koved, L. and Shneiderman, B. (1986) Embedded Menus: Selecting Items in Context. Communications of the ACM, 29, 312-318.

22. Benyon, D., Stone, D. and Woodroffe, M. (1997) Experience with Developing Multimedia Courseware for the World Wide Web: The Need for Better Tools and Clear Pedagogy. International Journal of Human-Computer Studies, 47, 197-218.

23. Thimbleby, H. (1997) Gentler: A Tool for Systematic Web Authoring. International Journal of Human-Computer Studies, 47, 139-168.

24. Erskine, L.E., Carter-Tod, D. and Burton, J.K. (1997) Dialogical Techniques for the Design of Web Sites. International Journal of Human-Computer Studies, 47, 169-196.

25. Smith, P.A., Newman, I.A. and Parks, L.M. (1997) Virtual Hierarchies and Virtual Networks: Some Lessons from Hypermedia Usability Research Applied to the World Wide Web. International Journal of Human-Computer Studies, 47, 67-95.

26. Bieber, M., Vitali, F., Ashman, H., Balasubramanian, V. and Oinas-Kukkonen, H. (1997) Fourth Generation Hypermedia: Some Missing Links for the World Wide Web. International Journal of Human-Computer Studies, 47, 31-66.

27. Lee, Y. and Kozar, K. (2012) Understanding of Website Usability: Specifying and Measuring Constructs and Their Relationships. Decision Support Systems, 52, 450-463.

28. Shneiderman, B. (1997) Designing Information-Abundant Web Sites: Issues and Recommendations. International Journal of Human-Computer Studies, 47, 5-30.

29. Buckingham, S. and McKnight, C. (1997) Word Wide Web Usability: Introduction to This Special Issue. International Journal of Human-Computer Studies, 47, 1-4.

30. Davis, P. and Shipman, F. (2011) Learning Usability Assessment Models for Web Sites. Proceedings of the 16th International Conference on Intelligent User Interfaces, Palo Alto, 13-16 February 2011, 195-204.

31. Nielsen, J. (2013) Usability 101: Introduction to Usability. Jakob Nielsen's Alertbox, 25.

32. Shneiderman, B., Kent, L. Norman, C.P., Benjamin, B., Bederson, D.A. and Golbeck, J. (2013) 30 Years at the University of Maryland's Human-Computer Interaction Lab (HCIL). Interactions, 20, 50-57.

33. Norman, D.A. (2010) Natural User Interfaces Are Not Natural. Interactions, 17, No. 3.

34. Myers, B.A. and Rosson, M.B. (1992) Survey on User Interface Programming. CHI'92 Proceedings of the SIGCHI Conference on Human Factors in Computing Systems, Monterey, CA, 3-7 May 1992, 195-202.

35. Landseadel, P. (1994) Methodologies in User Interface Design. IEEE Electronic Library, No. 0-7803-1910-9/94.

36. Nielsen, J. (1993) Usability Engineering. Academic Press, Inc., Harcourt Brace & Company, San Diego, USA.

37. Jacko, J.A. (2012) The human-Computer Interaction Handbook: Fundamentals, Evolving Technologies, and Emerging Applications. CRC Press, Florida, 1518.

CHAPTER 8

Code-reading support environment visualizing three fields and educational practice to understand nested loops

Koichi Yamashita[1*], Takamasa Nagao[2], Satoru Kogure[2], Yasuhiro Noguchi[3], Tatsuhiro Konishi[2] and Yukihiro Itoh[4]

[1]Faculty of Business Administration, Tokoha University, 1230 Miyakoda, Kita-ku, Hamamatsu, Shizuoka 431-2102, Japan.

[2]Graduate School of Informatics, Shizuoka University, 3-5-1 Johoku, Naka-ku, Hamamatsu, Shizuoka 432-8011, Japan.

[3]Faculty of Informatics, Shizuoka University, 3-5-1 Johoku, Naka-ku, Hamamatsu, Shizuoka 432-8011, Japan.

[4]Shizuoka University, 3-5-1 Johoku, Naka-ku, Hamamatsu, Shizuoka 432-8011, Japan.

ABSTRACT

In this paper, we describe a code-reading support environment and practical classroom applications using this environment to understand nested loops. Previously, we developed a code-reading support system based on visualization of the relationships among the program code, target domain world, and operations. We implemented the proposed system in exercises with nested loops. The evaluation results suggested that students could frequently fulfill learning objectives using the proposed system. However, we also discovered that some students experienced a learning impasse in the classroom. We attempted to address these students with two supporting approaches: bridging the gap between the generalization structures in the program code and their corresponding operations and enabling learners to predict the behavior of the nested loops. In this paper, we extend our previous system with new functions based on our two supporting approaches. Further, we implement the system in another classroom for nested loops. We describe a correlation between the proposed system and an understanding of nested loops using pre-/post-test comparisons. We discuss how code reading using the proposed system allows learners to cultivate a superior understanding of the program code.

KEYWORDS

Education for programming Domain world models Learning environment for exercise Classroom practice Learning by code reading

1. INTRODUCTION

With the continued rapid development and spread of information equipment such as smartphones, productivity improvements for program codes have been increasingly required. Software-developing environments have been enhanced, and programming languages have developed their descriptive capabilities. Programming skills are becoming necessary for not only software engineers. In the context of society informatization, increasing numbers of learners require an education in computer programming (Robins et al. 2003; Pears et al. 2007; Konecki & Petrlić 2014).

For several years, we have conducted programming classes for a wide variety of students. Through our experience in the classroom, we have paid attention to three fundamental skills that novice programming students tend to find difficult to acquire:

- Control structure can be and often is nested inside of another one. Students frequently have difficulty understanding nested structures in control flows and code descriptions. (Nesting)
- Students often struggle to generalize a set of concrete operations into an abstract function with variables. They tend to be able to interpret the processing contents of the first lap of the loop, the second lap of that, the third lap of that, and so on, while not to the contents of the *n*th lap of the loop. (Generalization)
- Students sometimes fail to grasp how the values of variables change with the execution of each statement. (Tracing; we treat tracing as tracing values of variables used in the code.)

Nested loops are a learning target with which novice learners frequently have an initial difficulty. This is because to fully comprehend this concept requires that the learner understands all three of the abovementioned fundamentals. Koppelman and van Dijk (2010) emphasized the importance of nested loops as one of the targets required to understand the concept of abstraction. However, limited exposure in programming courses constrains the efforts of learners to develop a thorough understanding of these fundamental concepts. The purpose of our study is to encourage students to learn these concepts efficiently. We have introduced learning support systems into classroom exercises in nested loops for several years (Kogure et al. 2013).

In this paper, we describe our code-reading support environment and the classroom applications of using this environment to understand nested loops. Previously, we developed a learning support system for code reading (Kogure et al. 2012). As in our previous work, we assume that learners will understand programs and algorithms by recalling an image consisting of three fields: the program code, objects processed by the program (i.e., the target domain world), and a sequence of concrete operations for the target domain. Learners must comprehend the relationships and correspondence among the components in each field. Many existing systems intuitively visualize the target domain world and reproduce the transition of its status. However, they do not have functions to visualize the correspondence between the program code and the concrete operations. The proposed system visualizes the three fields and their relationships to support understanding the relationships and correspondence among their components.

In our previous work (Kogure et al. 2013), we conducted exercise classes implementing our system to assist the understanding of nested loops and determined that some students experienced a learning impasse in the classroom. Based on the implicit and explicit feedback in our previous classroom practices, we constructed two approaches to cope with the impasse: bridging the gap between generalization structures in the program code and their corresponding operations and enabling learners to predict the behavior of a nested loop by observing two characteristics in the code.

In this paper, we present an overview of our previous system, new support strategies and functions that have been incorporated in the proposed system, and renewed classroom practice in helping students understand nested loops. We describe our previous learning support system, the two approaches to avoid the learning impasse, and the proposed extended system in the "Learning environment for programs and algorithms" section. We provide an overview of our classroom practice, our controlled experiment, and the evaluation results of the proposed system in the "Educational practice" section. Our evaluation results suggest that there is a correlation between our extended system and an understanding of nested loops. In the "Related works" section, we clarify the position of the proposed system, citing representative works. We discuss how code reading with the proposed system supports learners to cultivate an improved understanding of the program code and how the proposed system contributes to the overall learning of programming in the "Discussion" section. We conclude with a brief summary and discussion of future work in the "Conclusion" section.

2. LEARNING ENVIRONMENT FOR PROGRAMS AND ALGORITHMS

2.1 Our previous work

In our previous work (Kogure et al. 2012), we assumed that learners require an image consisting of three fields and understand the relationships among their components. The three fields are the program code field, target domain field, and operation field. With this assumption, we developed a system called learning environment for programs and algorithms (LEPA) that supported the learner in understanding programs. Figure 1 presents an overview of the learning support environment provided by LEPA. The three fields are reproduced in (a), (b), and (c).

LEPA is a code-reading support system for a typical program code in the C language in programming education for novice learners. Given a C code, unless the input code includes syntax errors and/or runtime errors, LEPA automatically generates a browser-based environment supporting learners to read the code. Consequently, it is not intended that learners use LEPA as a code-developing environment. LEPA provides learners with an environment to read an example code fragment provided by a teacher or textbook.

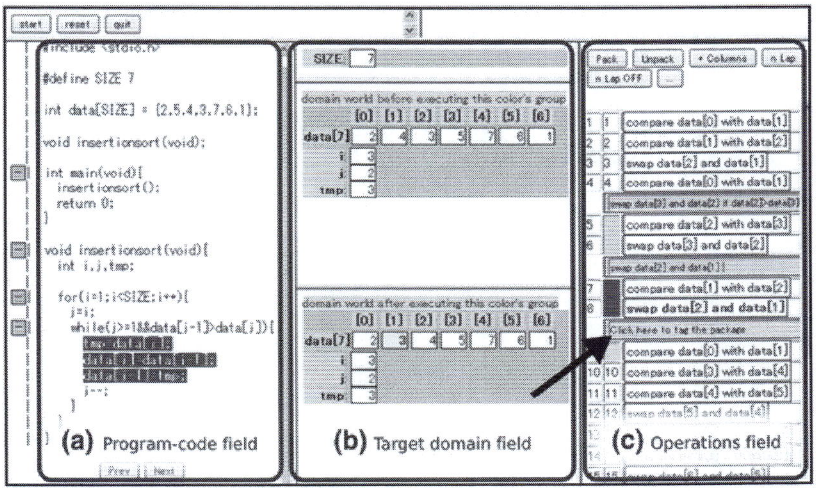

Figure 1. Overview of the LEPA environment, which has the program code field visualized in (**A**), target domain field in (**B**), and operation field in (**C**).

When a learner clicks on any of the operations in (c), the system displays the state of the target domain after executing the operation in (b). Thus, the learner can understand the role of a certain sequence of operations by comparing or observing the visualized target domain field before and after execution. Furthermore, the system highlights the code fragment in (a) corresponding to the selected operation in (c) and vice versa when the learner first clicks on a fragment in (a). The correspondence between a code fragment and an operation is therefore apparent to the learner. We believe that important information can be understood and retained using LEPA's visualization, including what happens in the target domain field when a concrete operation is executed and what kind of code is required to create this effect.

Learning with LEPA is based on the learner's externalization of what they have observed. Externalization is demonstrated with a process of packing and tagging using the GUI interface. The term "packing" in this paper means grouping a certain sequence of operations with single abstract function into a package. If such sequence of operations is discovered in (c), then the following two options are available:

1 The learner can push the "pack" button to pack the selected operation sequence into a package. Packages are permitted in a nested structure.

2 The learner can tag the package with a natural-language description according to its function (as indicated by the arrow in Fig. 1).

The resulting packed structure of the operation sequence ideally comes closer to the program code's structure. The learner will arrive at an understanding of the entire control sequence for the program code by completing a series of these activities.

Externalization activities can be classified into two groups according to the repetitiveness in the target operations: abstraction for operations that are not repetitive and generalization for operations that are repetitive. LEPA supports the former activity with packing operations and the latter by tagging the *n*th lap of a loop (as indicated in Fig. 2).

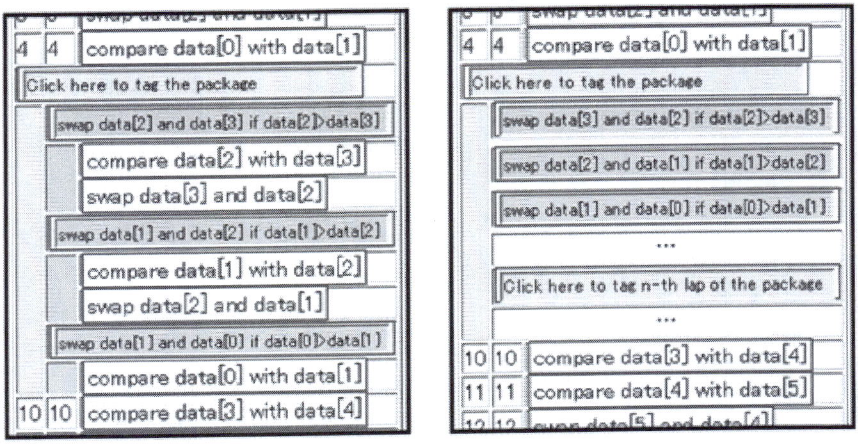

Figure 2. Tagging a package and tagging the *n*th lap of a package

2.2 Extending approach

In our previous work (Kogure et al. 2013), we implemented LEPA in exercise with nested loops. In these classes, we discovered that some students experienced a learning impasse. Based on the score differences between pre- and post-tests, the system did not indicate effects on the learner's ability to generalize. To address this impasse, we now extend LEPA based on two supporting approaches. First, we consider that a factor of the impasse was the gap between the generalization structures in the program code and in the corresponding operations. Therefore, we extend LEPA to support learners to bridge the gap. Then, we consider that a reduced ability to grasp the relationship between a code fragment and behavior hindered an improved understanding of nested loops. Therefore, we also extend LEPA to support learners to predict the behavior of a nested loop based on its classification derived by two characteristics in the code. In this section, we describe these two approaches. Hereafter, we call our previous system described in (Kogure et al. 2012) LEPA1 and the proposed extended system LEPA2.

2.2.1 Gap between the representations of generalization

Packing operations brings their structure closer to the program code. However, a gap can be found between the structure of a package and the corresponding fragment of the program code, as illustrated in Fig. 3.

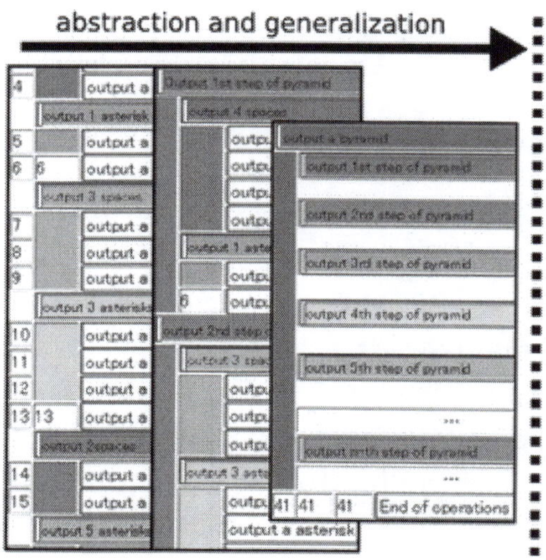


```
for(o=1; o<=N; o++) {
   for(i=1; i<=N-o; i++) {
      putchar(' ');
   }
   for(j=1; j<=2*o-1; j++) {
      putchar('*');
   }
   putchar('\n');
}
```

Figure 3 .Gap between the generalized operations and the program code

The right side in Fig. 3 presents the program code that displays an *N*-step pyramid by outputting an appropriate number of spaces, asterisks, and new lines. The left side indicates the transition of the package structure corresponding to the code. It is our opinion that this is the most plausible transition.

We assume that these packages are created with the following steps in terms of the learner's learning trajectory. First, the learner observes three fields and consequently creates two packages such as "output *x* spaces" and "output *y* asterisks." In this case, the learner assigns concrete numerals for *x* and *y*, rather than variables. Then, they are packed hierarchically with an operation that outputs a new line that is tagged "output *z*th step of the pyramid." At this stage, *z* is also a concrete numeral. The learner continues these steps until all of the operations in the operation field are included in the packages. Finally, *N* packages are acquired, packed, and tagged for the *n*th lap.

In LEPA1, the target for generalized tagging is only available at the first hierarchical level. This means that the learner tags the *n*th lap with variables that are exclusively from the package, "output *z*th step of the pyramid." LEPA1 hides any levels that are deeper than this in the operation field (i.e., outputting spaces, asterisks, and new lines). However, the program code for a nested loop is provided throughout the hierarchy. Having generalized the first-level hierarchy into "output *k*th step of the pyramid" with the variable *k*, the learner moves on to the deeper levels such as "output *N*-*k* spaces" or "output 2*k*-1 asterisks." Given generalizations throughout the hierarchy, the learner can realize the meaning of the control statements for the nested loops in the program code.

Based on these discussions, we have implemented a function that enables users to generalize a package while retaining the explicit hierarchy. Thus, the learner can tag the *n*th lap of the inner loops. Further, we have implemented a function that automatically generates a template for the general tag with the following steps:

1 Discriminating variable words from invariable words in the set of tags on the packages to be generalized

2 Replacing variable words with a series of symbols such as "_"

For example, in displaying a pyramid, the proposed system generates templates such as "output XXX spaces," "output ??? asterisks," and "output __th step of the pyramid" from the set of tags. By providing these templates, the proposed system encourages learners to formulate "XXX," "???," and "__" with loop control variables in the program code. The screenshot in Fig. 4 depicts the implemented functions. They are intended to assist learners to understand the structure of nested loops.

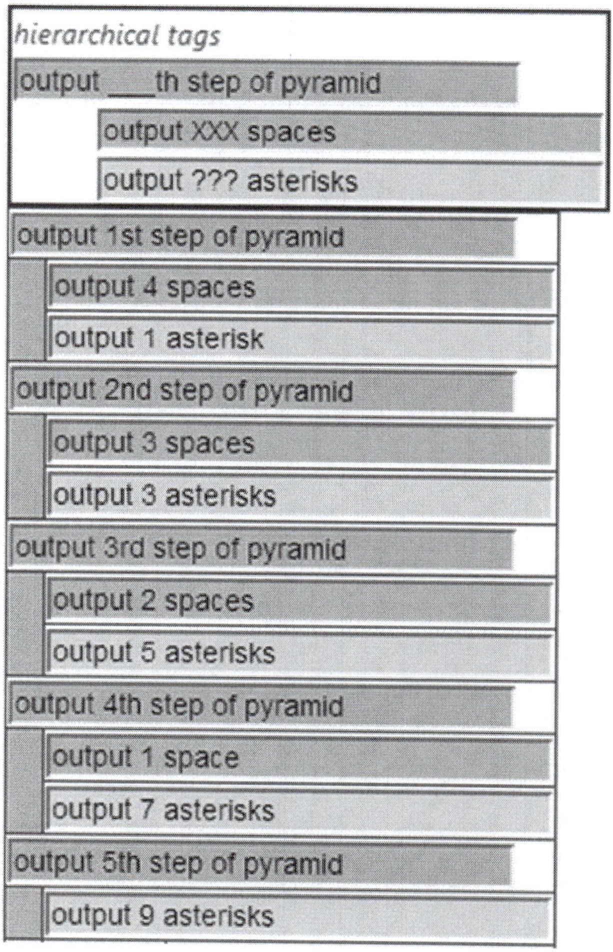

Figure 4 . Generalized packages while keeping the hierarchy explicit

2.2.2 Two characteristics in nested loop code

Programmers implement loops to include both identical operations and a wide variety of repetitive operations by appropriately referencing the loop's control variables. For example, different values are displayed by iterative execution of a statement to display the value of the variable *i* in a loop controlled by *i*. Therefore, the iteration package corresponding to the statement consists of a sequence of operations with different descriptions. Generalization in the proposed system involves two phases: identifying operations that regularly vary in the iterative package of the operation field and formulating this regularity. Nested loops increase the diversity of repetitiveness in the corresponding operations. We must consider not only the inner and outer loops where the control variables are referenced but also loops with references to variables that control the outer loop. This diversity can be regarded as a factor that novice programmers tend to find difficult to comprehend the relationship between the code of a nested loop and its behavior.

We constructed the second supporting approach to understand relationships by developing a classification of nested loops. By investigating textbooks and exercises, we have focused on the following two characteristics in nested loop code:

- Ch1. A statement to iterate in the inner loop references control variables from the outer loop.
- Ch2. A conditional statement in the inner loop references control variables from the outer loop.

A nested loop with Ch1 includes operations of varying representations at each iteration step of the outer loop—that is, for each inner loop. For example, the left side in Fig. 5 is the program code for outputting different values according to each iteration step of the outer loop. Conversely, a nested loop with Ch2 has a different number of operations corresponding to the inner loop according to each iteration step of the outer loop. For example, the right side in Fig. 5 is the program code where a different number of operations corresponds to the inner loop according to each iteration step of the outer loop.

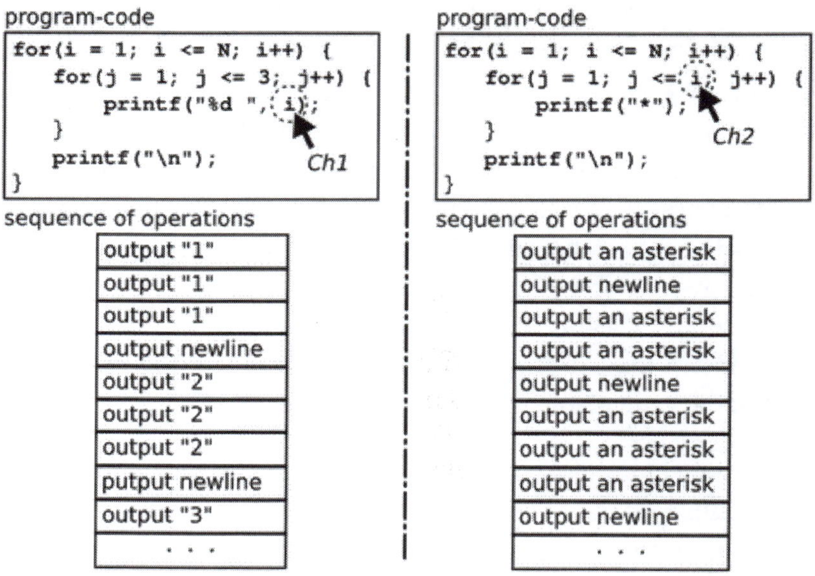

Figure 5. Program codes with Ch1/Ch2 and its operation sequence

We define "changing-step-contents type" as a type of nested loop with Ch1 and "changing-step-counts type" as a type with Ch2. The behavior of a nested loop can be classified into the following four classes based on the absence or presence of each characteristics:

- Without either Ch1 or Ch2—the operations corresponding to each inner loop and the number of them will never change.

- With Ch1—the operations corresponding to each inner loop will change in each outer loop step; however, the number will not.
- With Ch2—the number of operations corresponding to each inner loop will change in each outer loop step; however, the operations themselves will not.
- With both Ch1 and Ch2—both the operations corresponding to each inner loop and the number of them will change in each outer loop step.

In their packing series, learners must appropriately recognize the repetitiveness of an operation sequence as to whether it has these characteristics. Furthermore, to recognize the repetitiveness, learners must anticipate the regularity in the operation field based on the characteristics in the program code field. If learners can classify a nested loop code, they can reproduce the behavior of the nested loop based on the behavior pattern of the class. Moreover, by determining the behavior pattern of an algorithm in the coding, novice programmers can identify the characteristics to incorporate into the program code.

Based on these discussions, we attempt to teach students these characteristics, aiming to develop in them a better understanding of nested loops. Furthermore, we have implemented a function that involves requesting users to identify the type of nested loop provided in the program code field and whether it is the changing-step-contents type, the changing-step-counts type, both, or neither. For users who cannot provide an answer, LEPA2 suggests hints:

1 Hints related to where in the program code they should focus
2 Hints as to what characteristics they should be reading in the code revealed by the above hints
3 The characteristics that should have been read

2.3 Our extended system

In this subsection, we present an overview of the proposed extended system, LEPA2. Although the external appearance of LEPA2 is similar to that of LEPA1 (Fig. 1), LEPA2 places more emphasis on learning nested loops than LEPA1. LEPA2 helps learners to understand based on the following scenarios:

Ex1—*Tracing*: First, learners trace the program code and understand the behavior and control flow of the entire program. Figure 6 provides an overview of the environment constructed by LEPA2 at this phase. If the learners push the "Next" button, LEPA2 performs the following actions:

- Highlights the statement that will be executed next
- Displays the status of the target domain world after executing the statement
- Highlights the operation corresponding to the statement (if one exists)

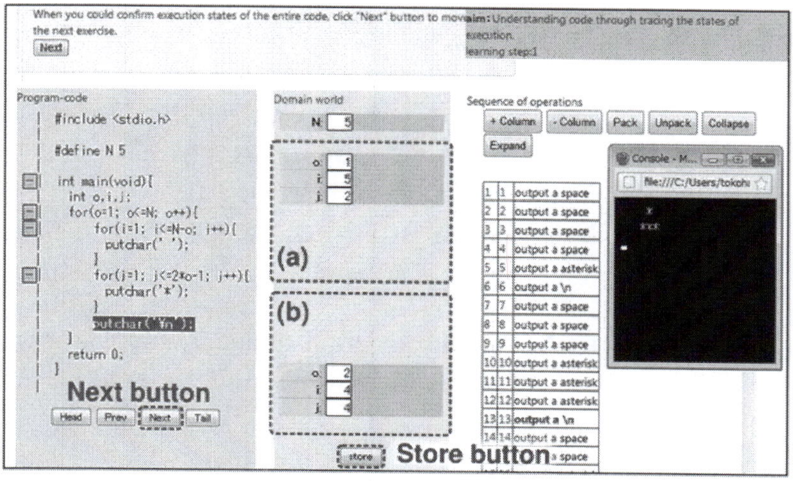

Figure 6. Ex1 on LEPA2

Moreover, if the learners click an operation in the operation field, LEPA2 highlights the statement in the program code field corresponding to the operation. If the learners push the "Store" button, LEPA2 copies the current status of the target domain world displaying in the area (b) into (a). This function support learners to observe the differences between the status of the target domain before and after executing specific statements or operations.

Ex2—*Abstraction*: Learners pack operations and recognize the abstract function of the package. Packing with nested structures helps learners to understand the nested structures of the entire program. Figure 7 provides an overview of environment constructed by LEPA2 at this phase. Learners find a certain sequence of operations that is making a change with a specific meaning in the target domain world, based on the findings in Ex1. The "Store" button is useful to fulfill this. Learners select the range of the sequence of operations in area (c) and pack the operations by clicking the "Pack" button. The resulting package is indented and put different colors on each other, and LEPA2 encourages learners to tag the package. Learners put the meaning of the changing into a natural-language phrase and input the phrase in the textbox set in the package.

Ex3—*Generalization*: After completing the packing for all the operations, learners generalize the packages corresponding to the nested loops, following the guidance of LEPA2. LEPA2 generates templates for the general tag based on the approach described in the "Gap between the representations of generalization" section. Learners complete the template according to the resemblance between the structure of the generalized package and that of the program code. Figure 8 provides an overview of the environment constructed by LEPA2 at this phase.

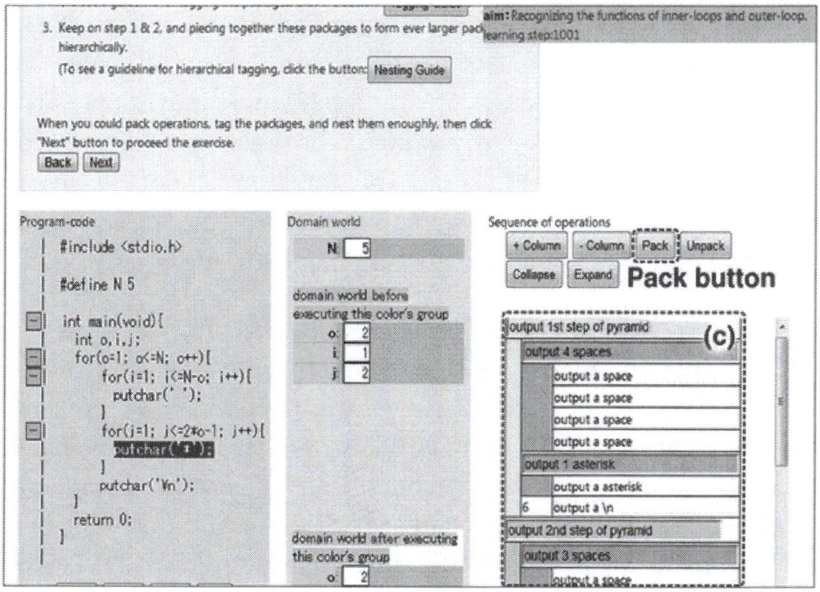

Figure 7 .Ex2 on LEPA2

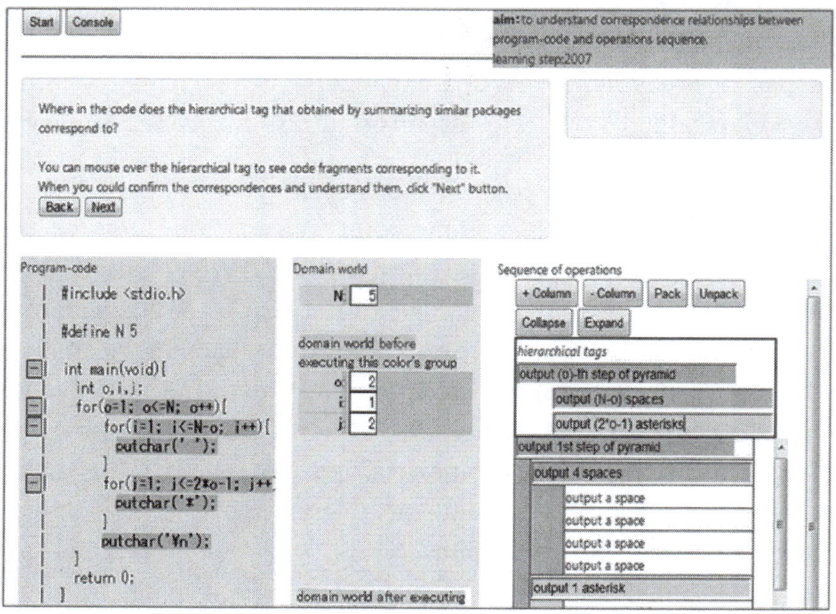

Figure 8 Ex3 on LEPA2

In nested loops, the repetitiveness to be discovered during the generalization phase does not appear in the concrete operations. Rather, it appears in the tags to packages packed by the learner. We have implemented the following functions to notify the learner in Ex3:

- Auto-complete function to support creating repetitive packages by displaying the learner's tagging history
- Function to support focusing on repetitive packages by displaying similar tags in the same color
- Function to support searching for the repetitiveness in the tags by hiding the concrete operations
- Function to support appropriately tagging the *n*th lap by providing a generalization template

Ex4—*Observing characteristics of the nested loop*: Lastly, learners observe the characteristics of the nested loop in the program code field to increase their understanding of nested loops. LEPA2 requests learners to identify the characteristics of nested loops based on the approach described in the "Two characteristics in nested loop code" section. Figure 9 provides an overview of the environment constructed by LEPA2 at this phase.

3. EDUCATIONAL PRACTICE

3.1 Hypotheses on learning effect with the proposed system

We conducted exercise classes to understand nested loops with LEPA2 as described in the previous section. Our previous educational practice (Kogure et al. 2013) suggested that learning with LEPA1 had the following effects:

- Learners could acquire the skills of tracing values of variables in the entire program code through observing the statuses in the target domain field.
- Learners could understand nested structures in control flows and code descriptions through packing sequences of operations hierarchically.
- Tags and structures of packages of operation sequences could visualize each learner's level of understanding and could provide information for better instructions to teachers.

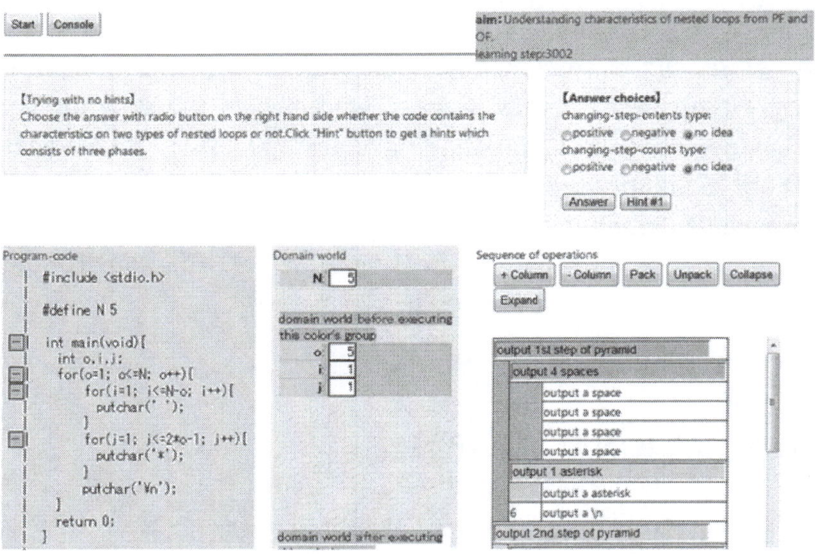

Figure 9 Ex4 on LEPA2

The exercises practiced in our previous work consisted of tracing the program code with nested loops and packing sequences of operations corresponding to the code. The learning scenario described in the "Our extended system" section includes the same tasks as the previous one, further tasks that generalize the packages corresponding to the nested loops based on the guidance of LEPA2, and tasks that observe the characteristics of the nested loop. Consequently, learning with LEPA2 is expected to help learners to cultivate better understandings of nested loops. Through the educational practice in this work, we evaluate the following learning effects for LEPA2:

- Hypo1. The redeveloped function of package generalization will promote an understanding of the behavior and structure of a nested loop.
- Hypo2. The function that requests an answer to a question based on the characteristics of the nested loop will promote an understanding of the relationships between the implementation and behavior of a nested loop.

3.2 Overview of the classroom practice

To verify the two hypotheses described above, we conducted an educational practice as a controlled experiment. We introduced LEPA2 in the experimental practice and did not introduce any learning support system in the controlled practice. There were 17 subjects in the entire experiment. In the experimental practice, two consecutive exercise classes were incorporated into a series of actual classes held in a university in Japan. The department holds two courses, Programming I and Programming II, for second-year students. Our experimental classes were conducted in the latter course. There were 12 participants in the experimental practice. All 12 participants were Business Administration majors, 20-year-old males, and had less than a year's experience in programming. In the control practice, two consecutive exercise classes were held out of the actual classes. There were five participants in the control practice. All five participants had the same properties as the experimental group, that is, they were Business Administration majors, 20-year-old males, and had less than a year's experience in programming. Table 1 presents a summary of the experimental and control classes.

Table 1. Summary of experimental and controlled classes

	Experimental	Controlled
Style of class	Lecture and exercise	Lecture and self-study
Learning time	90 min and after a week, 90 min	90 min and after 10 min, 90 min
Learning strategy	LEPA2	Traditional textbooks
The number of participants	12	5

For the experimental practice, the teacher who regularly taught the course lectured on single and nested loops for 1 h as a review in the first class. The lecture included an explanation of the characteristics of nested loops, as described in the "Two characteristics in nested loop code" section. At the end of the class, we conducted a 15-min pre-test to evaluate the students' understanding of nested loops before using LEPA2. In the second class, we allowed the students to use LEPA2 to learn nested loops. The program for this exercise displayed a five-step pyramid with spaces, asterisks, and new lines. Before the exercise in the class, the teacher described the aims of the exercise and the environment provided by LEPA2. During the exercise, neither the teacher nor our team provided assistance to the students for understanding the program. We recorded screen videos of the student's interactions with the environment. After the 60-min exercise, we conducted a 15-min post-test to evaluate the students' understanding of nested loops after using LEPA2. In the beginning of the first class, we informed that the students would have pre- and post-tests but the resultant scores of both test would not affect the course grades.

The programs used in the pre- and post-tests were different. However, the questions were almost identical. Question 1 (Q1) asked for the execution results expected from tracing the entire program by hand, including the nested loop. Question 2 (Q2) asked for the code fragment in the control of the inner loop with the execution results provided; question 3 (Q3) asked for the same for the outer loop. Question 4 (Q4) asked for the characteristics of the nested loops expected to appear in the program code, given the execution results only. Q1, Q2, and Q3 were designed to verify Hypo1; Q4 was designed to verify Hypo2. Table 2 presents a summary of each question in the pre- and post-tests.

Table 2. Summary of questions in pre- and post-tests

	Content of question	Maximum point
Q1	Describe the execution result of the program code.	3 (3 pts × 1)
Q2	Modify the control statement of the inner loop so as to output the indicated execution result.	9 (3 pts × 3)
Q3	Modify the control statement of the outer loop so as to output the indicated execution result.	3 (3 pts × 1)
Q4	What characteristics should be written in program code to output the indicated execution result.	48 (3 pts × 16)

For the control practice, students were first given the same 1-h lecture on nested loops as the experimental group, including an explanation of the characteristics of nested loops. We subsequently conducted the 15-min pre-test. Then, they studied nested loops using textbooks and the lecture material, without using LEPA2. Before studying, the teacher explained that the aim of the study was to understand nested loops using a sample program. The sample program was identical to the one in the experimental practice. The teacher also explained that, in particular, they should study tracing the program. After a 1-h study period, we conducted the 15-min post-test. Both pre- and post-tests were the same as those given to the experimental group.

Table 3 presents the differences in the average score for each question between the pre- and post-tests for both groups. We graded each question in the tests as follows: Q1 was worth three points; Q2 was worth nine; Q3 was worth three points; and Q4 was worth 48. There was a tendency for the experimental group's scores to improve significantly as indicated by the differences between pre- and post-tests, whereas in the control group they did not.

Table 3. The differences in the average scores between the pre- and post-tests

	Q1	Q2	Q3	Q4
Experimental	0.83	0.75	0.50	1.50
Control	1.20	0.20	−0.60	−5.70

We think that the improved results for the control group with regard to Q1 were a consequence of emphasizing tracing as the aim of the study. It is likely that the control group applied a significant portion of their time tracing. Therefore, we consider this group to have an advantage over the experimental group, whose study consisted of four stepwise exercises.

The experimental group had an interval of 1 week between pre- and post-tests because of schedule restriction. This was because we incorporated the classroom practice into a series of actual classes. This interval is not short and consequently allowed the experimental group to learn out of class with textbooks and course notes. Although the influences of this possibility on the score of the post-test must be considered, it is difficult to quantify them. However, we infer that the influences are practically trivial because all the experimental subjects had learned nested loops in Programming I classes held in an earlier semester. If the subjects had understood nested loops in that earlier class or by out-of-class learning, they would have tended to score high in the pre-test. Note that we conducted classes for nested loop again because the lack of student's understanding of nested loop hinders the exercise in Programming II. There was no actual tendency to score high; thus, we consider the influence of out-of-class learning as a negligible factor.

If we conducted both the pre- and post-tests in the same class session, the subjects would have a week interval between the pre-test and the classroom lecture because of the schedule restriction. In this case, the possibility of forgetting lecture contents would influence the score of the pre-test.

3.3 Evaluation based on learner's activities

We carefully reviewed the footage recorded from each student in the experimental group to analyze the learning effects in additional detail. Thereby, we found that the interactions with the learning environment provided by the proposed system differ significantly by student. For some activities made by using functions we expected to use, we classified the students into those who performed the activity (the positive group) and those who did not (the negative group). We rejected the activities that have greater than double times difference between the positive and negative groups to reduce the bias between the numbers of the both groups. Consequently, we focused on the following four activities:

- A1. The student consumed sufficient time (more than 3 min) tracing the program (Ex1).
- A2. The student packed all operation sequences corresponding to the inner loops and tagged all of the packages (Ex2).
- A3. The student formulated a general tag provided in the form of a template, using the guidance function from the outer loop generalization (Ex3).
- A4. The student used the function to observe the characteristics of nested loops (Ex4).

In A1, 3 min is the time we supposed to take to trace the entire program code with observing three fields. In actual exercise, there was bipolarization between the time to trace by the positive group, who take much more than 3 min, and the time to trace by the negative group, who take less than 1 min.

We consider these activities to lead to the following learning effects:

- A1 advances an understanding of the behavior of an entire nested loop from which the student will acquire tracing skills.
- A2 promotes an understanding of the step contents in the inner loops from which the student will understand the structure of a nested loop.
- A3 assists the understanding of the step contents in the outer loop from which the student will acquire the skills required to generalize concrete operations.
- A4 leads to an understanding of the characteristics of nested loops from which the student will understand the relationships between the behavior and the characteristics in the program code.

We expect that students performing these activities will improve their marks as follows: students who performed A1 will score better on Q1; those who performed A2 will score better on Q2; and likewise for those performing A3 and A4 for Q3 and Q4, respectively.

Table 4 presents the differences in the average scores for each question between the pre- and post-tests in the positive and negative groups for A1 through A4. By performing an independent *t* test, if the difference between the negative and positive groups is statistically significant such that the progress level reached $p = 0.05$, we placed an asterisk next to the number. As expected, each positive group demonstrated significant progress for their marks on the corresponding question.

Table 4. Differences in the average scores for positive and negative groups

		Number of students	Q1	Q2	Q3	Q4
A1	Positive	7	1.43*	1.57	1.29*	
	Negative	5	0.00	−0.50	−0.75	
A2	Positive	5	0.80	2.60*	0.60	
	Negative	7	1.00	−0.67	0.50	
A3	Positive	4	1.17	1.67	1.50*	
	Negative	8	0.60	−0.20	−0.60	
A4	Positive	4				2.63*
	Negative	8				−0.75

A1 to A4 are the four learning activities; $^*p < 0.05$

We must consider that a statistically insufficient number of the subjects may influence the accuracy of verification. Our evaluation results do not have sufficient reliability because we could not procure a sufficient number of subjects in the classroom practice. The course we selected for our class is optional for the faculty, and the number of applicants varies from year to year. Whereas practice in actual classes provides practical learning information, the number of subjects is related to the number of applicants. However, we believe that continuous practice will suppress this matter. The progress with regard to Q1, Q2, and Q3 suggests in favor of Hypo1 and that of Q4 for Hypo2. Hence, although preliminarily, the results suggests that the proposed system results in an improved understanding of nested loops provided the user performs the expected activities. We must consider that a statistically insufficient number of the subjects may influence the accuracy of verification. However, we believe that continuous practice will suppress this matter.

4. RELATED WORKS

In general, programming students learn algorithm, code readings, and coding in turn. They attend a lecture and receive algorithm instruction from their teacher. Then, they reproduce the behavior of the algorithm using certain input data and produce a sequence of concrete operations that represents the behavior of the algorithm. Subsequently, they abstract sequences of operations, grasp the relationship between the abstracted operations and the program code, and consequently understand the entire program code. Finally, they perform a coding exercise to confirm their understandings.

Thus far, several intelligent tutoring systems have been developed to support programming learners. These include RoboProf (Daly & Horgan, 2004), JITS (Sykes & Franek, 2003), J-LATTE (Holland et al. 2003), and BITS (Butz et al. 2006). Moreover, several learning support systems based on visualizing algorithms have received attention, including TRAKLA2 (Malmi et al. 2004), Jeliot 3 (Moreno et al. 2004; Čisar et al. 2011), and ViLLE (Rajala et al. 2008). These systems can be classified from the standpoint of the tasks required for understanding an algorithm or program code and tend to support one or the other. We believe that an attractive learning target can be found in the gap between these two tasks. These systems, however, do not offer a suitable means for bridging the gap.

Learners who have a proper understanding of an algorithm can reproduce its behavior with concrete data. A sequence of operations in LEPA is a sequence of natural-language descriptions representing the algorithm's behavior. Hence, the operation sequence can be regarded as an externalization of the learner's understanding of the algorithm. Other existing systems visualize the relationship between the program code and its target domain world. LEPA provides this as well and furthermore visualizes the relationship between a sequence of operations and its target domain. It also visualizes the correspondence of a code fragment to its operation. We expect that the visualization of these three fields and the relationships among them contributes to bridging the gap between the two tasks.

When a program-comprehension task is assigned to a programmer, the procedure for reading the code is normally a dual process: first, recognizing the function of groups of statements and then piecing together these chunks to form ever-larger chunks (Shneiderman & Mayer, 1979). Programmers proceed through these steps hierarchically until the entire program is understood. LEPA offers a function to support learners in their endeavor to envisage behavior similar to that in the operation field. Learners can learn an entire series of code-reading process using LEPA by this function.

Recently, learning support systems with an integrated development environment have been developed, aiming to support the entire programming exercise (Đanić et al. 2011; Gerdes et al. 2012; Neve et al. 2012). These systems do not focus on the code-reading stage. Consequently the learners only externalize their understanding of the algorithm and program code by coding. Learners and teachers cannot determine the cause of a learner's lack of understanding: lack of algorithm understanding, program code understanding, or coding skills. Further, we can identify systems based on externalization with other methods than coding (Cooper et al. 2012). However, they are intended to support younger learners and cannot directly support users to read and write the program code. More discussions are required to introduce these systems into current programming education at the university level.

In contrast, LEPA encourages the learner to abstract groups of statements in the program code and to externalize the abstracts in the form of tags. Learners trace the program code, observe the changes in the target domain field, pack the operations in the operation field according to the observation, and tag the packages according to the function. The learner's package structure is the product of externalization with a series of these activities. LEPA does not include correct package structure solutions, hence never provides a solution, and never leads the learner to the correct solution. Thus, teachers can consider the package structures as the learner's understandings externalized by the proposed system. As will be shown in the next section, we consider that these functions improve the quality of programming education.

There are systems available that target nested loops, including AlgoTutor (Yoo et al. 2012) and the tutoring system developed by Dancik and Kumar (2003). However, it is not clear how they lead learners to understand or acquire the fundamental concepts described in the "Introduction" section. If any chunk corresponds to an iteration such as a loop, it is often the case that the learner's recognition involves generalizing with variables. LEPA2 bridges the gap between hierarchical structures of chunks in a sequence of operations and those of descriptions in the program code and aims to improve the learning supports for nested loops by elaborating the strategy of a hierarchical procedure for reading code.

5. DISCUSSION

In this section, we discuss the contribution of the proposed system to learning programming. The correlations suggested by the classroom practices described in the "Educational practice" section and in our previous work (Kogure et al. 2013) can be summarized as follows:

- Learning with the proposed system can contribute to acquiring fundamental skills to understand programs.
- Tags and structures of packages of operation sequence can visualize each learner's level of understanding and can provide information for better instruction to teachers.

In traditional education for programming in the code-reading stage, the teacher explains the functions and the behavior of the sample code in the textbooks. Learners require a detailed understanding of the sample code because it is typical of each learning target. However, code reading is a difficult learning activity to externalize achievements. Teachers attempt to understand indirectly the learner's level of understanding by exercises in writing similar code or tracing values of the variables. It is not obvious that the achievements of these exercises reflect appropriately the level of understanding of the sample code.

As described in the previous section, learners with the proposed system externalize their understandings in observable forms for themselves and their teachers through packing, generalizing, and tagging operations in the operation field. To confirm the validity of the proposed system's externalization, we invited six students to read code with the proposed system as a pilot experiment. The students had not attended the practices and experiments described in the "Educational practice" section. The code was from an early section of our textbooks. Figure 10 presents the code and the corresponding sequence of operations. Figure 11 presents the packages generated by the students. There are several observations of the students' understandings:

- Some students packed the operations assigning constants and operations assigning calculated values separately, whereas others combined them into a single package.
- Some students recognized that data is stored to a letter rather than a variable.
- Some students recognized the settings of the values into variables *a*, *b*, and *c* as calculations rather than assignments.

The intent of the sample code was to calculate and display the value 10 + 20. A lack of understanding of intention could possibly yield the first and third developments. Students with recognition as in the second development could possibly not appropriately grasp the concept of variables.

program-code

```
#include <stdio.h>

int main(void)
{
    int a, b, c;

    a = 10;
    b = 20;
    c = a + b;
    printf("%d\n", c);

    return 0;
}
```

sequence of operations

assign *10* to *a*
assign *20* to *b*
assign *a* + *b* to *c*
display the value of *c*
end of operations

Figure 10 .Program code from textbooks and its operation sequence

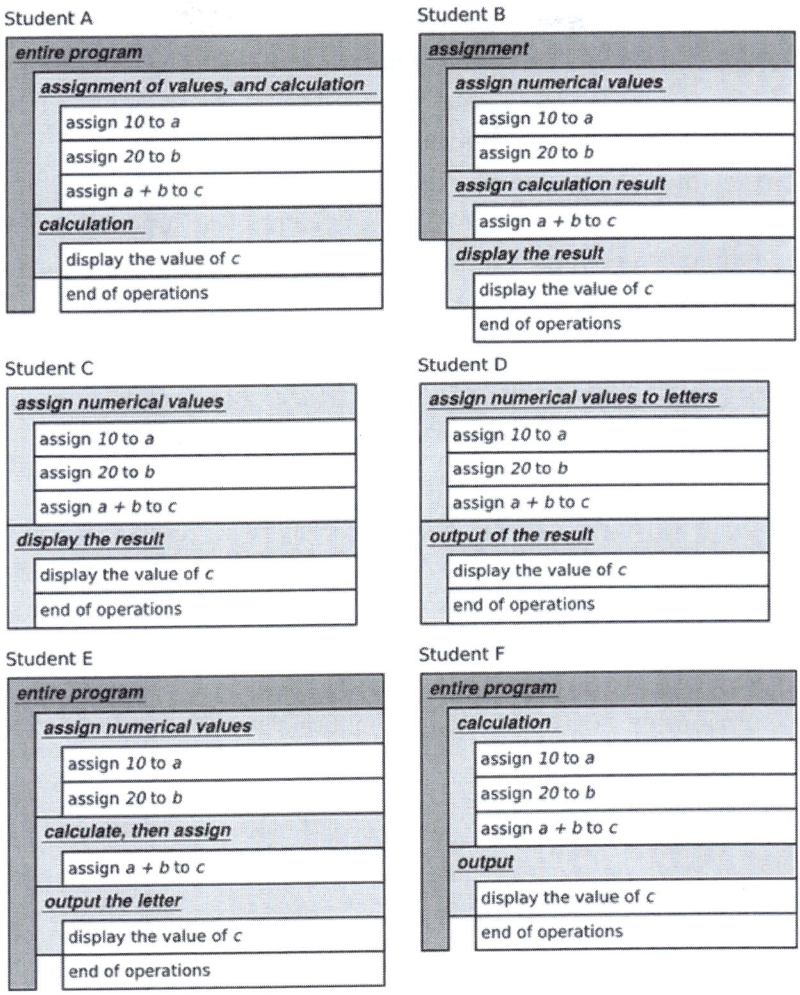

Figure 11. Package structures generated by six students

We consider that the proposed system improves the quality of programming education based on these observations. In the classroom exercise with the proposed system, teachers can observe the status of the students' code reading by walking around the class. Because the students are externalizing their understanding using the proposed system, the teachers can grasp the students' understandings by observing operation field in each student's environment. If there is a misunderstanding regarding package structures and tags generated by a certain student, the teacher can offer the student high-quality instructions by separate interaction. If the teacher identifies a tendency of the class to misunderstand as a whole, they can provide the supplemental instruction in the form of classroom lectures. Enabling teachers to choose the appropriate style of instruction improves the quality of education and leads students to a superior understanding of the program code.

For example, in the results of the pilot experiment, there is a student (student D) suspected of not appropriately understanding the concept of variables. There is also a tendency of the class not to distinguish between using a constant number from reference variables. This information provided by LEPA allows teachers to make a decision as to the required instruction, especially to make opportunity discovery. In the pilot experiment, we did not evaluate the contribution degree. We plan to conduct further experiments to obtain more information regarding this.

In programming education, there are minimal reports available on the learning effect of code reading. In our opinion, this is caused by the difficulty in externalizing learning outcomes. As indicated in Fig. 11, package structures in the proposed environment provide useful information to teachers regarding a learner's understanding. Thus, we consider that the proposed system not only directly supports learners to cultivate a superior understanding of the program code but also indirectly by improving teacher's performance.

6. CONCLUSION

In this paper, we described our learning support system and classroom practices with the proposed system for an improved understanding of nested loops. Nested loops are an appropriate target for learning the fundamental skills of programming. We believe that learning support systems efficiently and effectively contribute to an understanding of fundamental programming concepts and the acquisition of the skills required to utilize them.

In the classroom practice with LEPA1, we determined that some students had experienced a learning impasse. We attempted to address this with two supporting approaches: bridging the gap between generalization structures in the program code and their corresponding operations and enabling learners to predict the behavior of nested loops by observing two characteristics in the code. For the former approach, we developed a function for generalizing packages by maintaining an explicit hierarchy. For the latter, a function allowing learners to observe the characteristics of nested loops was provided.

We conducted a controlled experiment to evaluate in detail the effect of LEPA2 with the abovementioned functions incorporated. In the experimental practice with 12 students forming an experimental group, we incorporated LEPA2 in an actual classroom. In the controlled experiment with five students forming a control group, we evaluated learning effects using traditional teaching materials such as textbooks. The evaluation results based on the scores calculated from a pre- and post-tests suggest that there was a correlation between using LEPA2 and an understanding of nested loops although preliminarily.

Package structures and tags generated in learning with the proposed system can be regarded as an externalized learner understanding that is observable by teachers. Feedback provided by these is useful information for teachers' instruction. Consequently, qualitative improvements of instruction are expected. Hence, we consider that the proposed system directly and indirectly supports learners to read the program code and cultivate a better understanding of the program code.

We can highlight two limitations of the proposed works derived from classroom practices incorporated into a series of actual classes. First, our evaluation results do not have sufficient reliability because we could not procure a sufficient number of subjects in the classroom practice, as described in the "Evaluation based on learner's activities" section. We expect that continuous practice will overcome this matter. We plan to continue to conduct classroom practices using the proposed system in the future.

Furthermore, we must allow a 1-week interval between pre- and post-tests in experimental practice because of the schedule restriction. We infer that the influences of this are minimal, as described in the "Overview of the classroom practice" section. It is difficult or almost impossible to quantify and eliminate these factors. We continue to deliberate from various viewpoints such as conducting educational practice outside of actual classes and constructing an efficient learning scenario where we can evaluate learning effects more accurately during a class.

In future work, we will consider how the proposed system contributes to improve education for programming, as described in the "Discussion" section. We will collect package structures from the operation fields packed by learners through further classroom practice and evaluate the relationships between the students' understandings and those structures. Moreover, we will develop an evaluation method to measure how the proposed system contributes to the decision-making on teachers' instruction. Furthermore, we will continue to develop a desirable learning scenario that naturally causes the learners to perform the appropriate learning activities expected by the proposed system.

ACKNOWLEDGEMENTS

This study was supported by Japanese Grant-in-Aid for Scientific Research (B) 24300282.

COMPETING INTERESTS

The authors declaire that they have no competing interests.

AUTHORS' CONTRIBUTIONS

KY participated in the development of the system, conducted the classroom practices, summarized the whole research and wrote this paper. TN and SK implemented and evaluated the system. YN, TK and YI gave advices based on actual teaching experience. All authors read and approved the final manuscript.

REFERENCES

1. Butz, C. J., Hua, S., & Maguire, B. R. (2006). A web-based Bayesian intelligent tutoring system for computer programming. Journal of Web Intelligence and Agent Systems, 4(1), 77–97.
2. Čisar, S. M., Pinter, R., Radosav, D., & Čisar, P. (2011). Effectiveness of program visualization in learning java: a case

study with Jeliot 3. International Journal of Computer, Communications & Control, 6(4), 669–682.

3. Cooper, S., Jae Nam, Y., & Si, L. (2012). Initial results of using an intelligent tutoring system with Alice. In Proc. of the 17th ACM annual conference on Innovation and technology in computer science education, 138–143

4. Daly, C., & Horgan, J. M. (2004). An automated learning system for Java programming. IEEE Transaction on Education, 47(1), 10–17.

5. Dancik, G., & Kumar, A. (2003). A tutor for counter-controlled loop concepts and its evaluation. Frontiers in Education Conference, 1(T3C), 7–12.

6. Đanić, M., Radošević, D., & Orehovački, T. (2011). Evaluation of student programming assignments in online environment. In Proc. of the 22nd Centerl European Conference on Information and Intelligent Systems, 111–116

7. Gerdes, A., Jeuring, J., & Heeren, B. (2012). An interactive functional programming tutor. In Proc. of the 17th ACM Annual Conference on Innovaion and Technology in Computer Science Education, 250–255

8. Holland, J., Mitrovic, A., & Martin, B. (2003). J-LATTE: a constraint-based tutor for Java. In Proc. of the 17th Int. Conference on Computers in Education, 142–146

9. Kogure, S., Okamoto, M., Noguchi, Y., Konishi, T., & Itoh, Y. (2012). Adapting guidance and externalization support features to program and algorithm learning support environment. In Proc. of the 20th Int. Conference of Computers in Education, 321–323.

10. Kogure, S., Okamoto, M., Yamashita, K., Noguchi, Y., Konishi, T., & Itoh, Y. (2013). Evaluation of an algorithm and programming

learning support features to program and algorithm learning support environment. In Proc. of the 21st Int. Conference of Computers in Education, 418–424.

11. Konecki, M., & Petrlić, M. (2014). Main problems of programming novices and the right course of action. In Proc. of the 25th Central European Conference on Information and Intelligent Systems, 116–123.

12. Koppelman, H., & van Dijk, B. (2010). Teaching abstraction in introductory courses. In Proc. of the 15th Annual Conference on Innovation and Technology in Computer Science, 174–178.

13. Malmi, L., Karavirta, V., Korhonen, A., Nikander, J., Seppälä, O., & Silvasti, P. (2004). Visual algorithm simulation exercise system with automatic assessment: TRAKLA2. Informatics in Education, 3(2), 267–288.

14. Moreno, A., Myller, N., Sutinen, E., & Ben-Ari, M. (2004). Visualizing programs with Jeliot 3. In Proc. of the Working Conference on Advanced Visual Interfaces, 373–376.

15. Neve, P., Gunter, G., Livingstone, D., & Orwell, J. (2012). NoobLab: an intelligent learning environment for teaching programming. In Proc. of the 2012 IEEE/WIC/ACM Joint Conferences on Web Intelligence and Intelligent Agent Technology, 3, 357–361.

16. Pears, A., Seidman, S., Malmi, L., Mannila, L., Adams, E., Bennedsen, J., et al. (2007). A survey of literature on the teaching of introductory programming. ACM SIGCSE Bulletin, 39(4), 204–223.

17. Rajala, T., Laakso, M.-J., Kaila, E., & Salakoski, T. (2008). Effectiveness of program visualization: a case study with the ViLLE tool. Journal of Information Technology Education, 7, 15–32.

18. Robins, A., Rountree, J., & Rountree, N. (2003). Learning and teaching programming: a review and discussion. Computer Science Education, 13(2), 137–172.

19. Shneiderman, B., & Mayer, R. (1979). Syntactic/semantic interactions in programmer behavior: a model and experimental results. International Journal of Computer and Information Sciences, 8(3), 219–238.

20. Sykes, E. R., & Franek, F. (2003). An intelligent tutoring system prototype for learning to program Java. In Proc. of the 3rd IEEE Int. Conference on Advanced Learning Technologies, 485–492.

21. Yoo, J., Yoo, S., Seo, S., Dong, Z., & Pettey, C. (2012). Can we teach algorithm development skills? In Proc. of the 50th Annual Southeast Regional Conference, 101–105.

Index

B

Bayesian estimation, 68, 80
Bayesian Network, 1, 4, 5, 12
Beijing, 67, 68, 72, 73, 85, 86, 88, 96, 97, 98, 100, 103, 107, 108, 109, 111
Bias, 50, 52

C

Climate Change, 189, 190, 201, 207, 212, 213, 215, 216
conditional autoregressive model, 68

D

Dummy coding, 35

E

Ergonomics, 218
estimación bayesiana, 68

G

Growth models, 141, 142, 146, 149, 150, 154, 159, 170, 177

H

HCI, 218, 219, 220

I

Imputation, 29, 37, 39, 61, 63
Internet interventions, 142, 169

L

Large-scale assessment, 29, 30
Learning by code reading, 250

M

MCAR Models, 78
Mediation, 142, 183
Missing data, 29, 32, 34, 61, 63, 163, 181, 186
Mixture Models, 114
MMI method, 29, 42, 47, 55, 56
modelado a nivel múltiple, 68
modelo de auto-regresión condicional, 68
Moderation, 142, 170
multilevel modeling, 67, 68, 93, 99, 101, 102, 106, 174, 176, 182
Multilevel modeling, 108, 142

O

Ontology Schema, 114

R

Reputational Risk, 114
response theory, 29, 64
RMSE, 30, 48, 49, 50, 53, 54, 56

S

Simple Climate Models, 190
Simulation Item, 29
SMI method, 47, 56
Statistical Models, 113, 139, 217, 218, 243
Statistical-Dynamical Models, 189, 190, 206, 209
Structural equation modeling, 142
subjective travel satisfaction, 68, 72, 73, 85, 86, 88, 96, 103

T

Textual Data Analysis, 114

U

Usability, 217, 218, 219, 230, 242, 244, 245, 246, 247

W

Web Pages Interface, 218